Preface

In keeping with the ideas set out by the Stuart International Model Engineers' Club, this book is offered in the hope that it will help newcomers to the craft of building model steam engines.

Although these engines are described as models, they are, in fact, real power units, and all the normal engineering machining and fitting processes are required for their completion. To tackle the building of such an engine from a set of castings, requires the will to accomplish, and the resourcefulness to make the fullest use of limited and simple equipment.

Complete novices, with no workshop equipment, may be interested in the engines, requiring assembly only, which are listed in the Stuart Turner catalogue.

To keep this book to a reasonable length, it has been presumed that anyone tackling an engine of this size is not completely without workshop experience; for this reason, the finer details of tool shapes, tapping drill sizes, speeds and feeds, etc., have been left to the readers' own experience. Where this may be lacking, the writer would recommend the range of workshop handbooks issued by "Model and Allied Publications Ltd.," 13/35 Bridge Street, Hemel Hempstead, Herts. In addition, a perusal of the SIMEC List of Members may well produce the address of a nearby enthusiast willing to give reasonable assistance.

Finally, may I sincerely thank Tom Nevins, the Secretary of SIMEC, for suggesting this book; and Stuart Turner Ltd., without whom model engineering would be a much less interesting pastime.

ANDREW SMITH.

Saltford, Avon, England.

ISBN 0 905180 00 3.

Published by Henley Publications.
Reprinted December 1975.

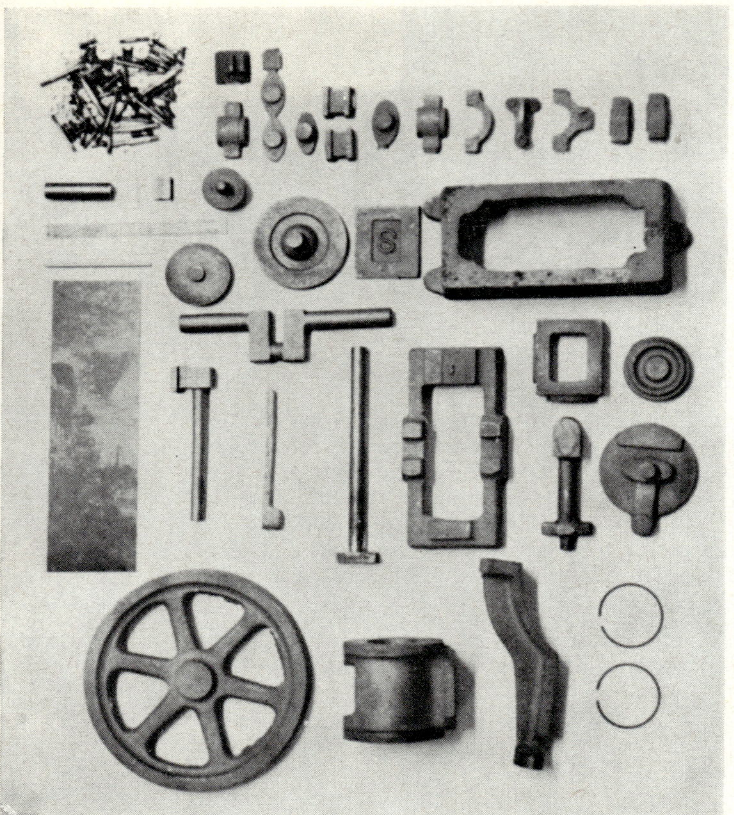

The full set of castings and materials for the No. 1 Engine, as supplied by Messrs. Stuart Turner Ltd.

CONTENTS

EDITOR'S NOTE

All the diagrams and figures referred to in the text will be found at the end of the booklet, on detachable pages, so that they can be removed, if desired, and used alongside the relevant page of text, or pinned up over the work-bench for convenience. The item numbers refer to Stuart Drg. No. 90007.

T.H.W.N.

During the final decade of the last century, one wonders whether Sidney Marmaduke Stuart Turner could possibly have imagined that, over three-quarters of a century later, a steam enthusiasts would still be purchasing sets of castings and drawings of the model vertical steam engine which he had just then designed, built and begun to market. It cannot be very often that the product on which a business is started has been so well conceived that eighty years later it is still a major item in the sales catalogues of the firm. During all this time, the building of the Stuart Turner No. 1 Engine must have given pleasure to many thousands of model engineers and steam enthusiasts all over the world and I sincerely hope that perhaps these words of mine may reintroduce this engine to newcomers in this field.

I hesitate to describe this engine as a model. Although it is certainly a very close scale replica of the type of power unit which drove electric generators, pumps and workshop machinery at the turn of the century, nevertheless with its bore and stroke of 2 inches and sturdy construction, it is capable of doing useful work if so required.

In accepting the pleasurable task of carrying through and describing the construction of this engine, I have borne in mind the limitations in skill and equipment of the average beginner of whatever age. A lathe is, of course, a necessity—fitted with a back-gear, but such refinements as screwcutting and self-feed are not really needed. The usual Myford ML7 is obviously an ideal choice, but I hasten to point out that in fact the lathe used was of the light $3\frac{1}{2}$ inch type which could be bought for a few pounds many years ago. Such lathes were sold under many different trade names but were frequently made by the old established Sheffield firm of Portass. An engineer's vice of 3 or 4 inch jaws and—most important—firmly mounted on a substantial bench. Set the vice jaws so that they are just level with your elbow and filing will lose most of its backache.

Although the lathe is perfectly suitable for drilling, a drilling machine greatly facilitates this work. The machine in my workshop is at least eighty years old; has a throat depth of over 18 inches and, I am told, is of a type used in the early days of bicycling for wheel building. Whether true or not it is certainly a substantial bench machine which I have fitted with electric motor and geared chuck. With simple facilities for tool and drill grinding we are ready to begin. I have said nothing about small tools but these will become evident as we progress.

Perhaps, like me, you are of an age which thought that the child-like pleasure of unpacking a large parcel belonged to the dim past. No so! When the wire bound package arrives bearing the Stuart Turner label you will find that the long forgotten tingle of excitement is still there. By the way, save the wire, you will find it very useful in a number of machining processes!

Check each item carefully against the Materials Schedule, including each screw, stud and nut. Lay out each casting and piece of material alongside its drawing. Many of the details are drawn half or quarter size and it is only when you handle the castings that you realise you are dealing with some quite hefty pieces of cast-iron.

Where to start? There are many ideas on this. If you are well equipped with precision measuring equipment you may decide to commence with Item 1, make each part to precision measurements and finally assemble them all in one glorious flourish of skill! But if your measuring kit consists of a 12 inch steel rule, trysquare, calipers and perhaps a 0-1 inch micrometer, then to start at the bottom and progressively make and erect, checking each part against its neighbour and using dimensions as a guide rather than a necessity, may be the best policy. After all, this is probably the way in which this engine, both model and full-size, was originally made! This is the technique we will use.

BOX BED

When I showed the box bed casting to my small lathe with the intention of machining the top and bottom surfaces, I'll swear that the lathe shrank at least three inches! The casting requires a gap which will swing 9 inches diameter and $2\frac{1}{2}$ inches wide. However, a few minutes checking with the steel rule as a straight edge and it was obvious that all that was needed was a little work with a flat file. Check where the high spots, if any, are and carefully work them down. The main thing is to avoid rounding the surface. Do not remove more than you have absolutely got to in order to get a flat seating (a) for the box bed on its floor and (b) for the engine soleplate on the top of the box bed. Half an hour should see the job complete with 17/64th inch holes drilled in the three bolting down lugs; but do spend some little time cleaning up the sides to help with the paint finish later. Figure 1 shows the general dimensions of the boxbed.

SOLE PLATE—Item 9

This item will use the many capabilities of the lathe, but first the underside must be made smooth and true. I found it was quicker to do this with a file than waste time setting the casting up in the four-jaw chuck or on the face-plate. Check the flatness of the sole plate by testing for rock on the lathe bed or other accurate flat surface. When the underside is true, mount the soleplate in the four-jaw chuck or clamp on the faceplate as shown in Fig 2. The tops of the bearing housings may then be faced

off to give a height, such that the depth of the gap to take the bearing brasses is just slightly less than $\frac{7}{16}$ inch. Feed the tool forward $\frac{3}{16}$ inch and face the seating for the standard (Item 6). Ideally the seating for the turned column (Item 11) should not be touched. This seating is, as cast, already at the required angle and, hopefully, it will only require a few minutes' work with a file to prepare it for the bottom of the turned column.

At this point it is worth stressing that all machined surfaces on castings should be machined with careful reference to the surfaces which are to be left unmachined. A typical example of this is a flywheel whose periphery has been machined without reference to the rest of the cast shape. The eccentricity as it rotates not only looks hideous, it is also positively dangerous.

Equal amounts should now be removed by filing from the inner surfaces of the soleplate to give the required width of one and fifteen sixteenths between the bearing pedestals. The positions of the four 17/64 inch holes for the holding down bolts are now marked out and drilled. Check the castings carefully as I found that my holes had to be at $2\frac{1}{8}$ inch centres, not as shown on the drawing, so that the $\frac{1}{2}$ inch diameter spot-facings would not foul the seating for the standard. Drill all holes with a pilot drill first before drilling to size. Spot-facing cutters, commonly called 'pin-drills', may be obtained from most suppliers of model engineering equipment including Stuart Turners.

At this stage in the proceedings the soleplate might well be left while the main bearings (Item 5) are completed. However, in order to keep this description clear we will continue the written word with the machining of the soleplate, but, do please appreciate that, in practice, you will require the finished main bearings in order to check the accuracy of the soleplate bearing housings.

The assembly of boxbed, soleplate, crankshaft, main bearings and flywheel after the first part of the work has been completed.

To face the housings to $\frac{9}{16}$ inch thick an end mill of at least this diameter will be required. This can be made, without any great difficulty, from a short piece of $\frac{5}{8}$ or $\frac{3}{4}$ inch silver steel, see Fig. 3. Mount it well back in the three-jaw chuck when in use to give ample support. Clamp the soleplate to the cross-slide with suitable packing so that the bottom of the machined face will be at the correct distance from the top of the bearing housing.

If your lathe is rather light, or perhaps you are just nervous, spend some time filing most of the unwanted metal away leaving only about $\frac{1}{32}$ inch to be removed by the milling cutter. I practise this policy whenever possible. It maintains and enhances your skill with a file; if it is planned as the first job on entering the workshop it saves switching on an electric fire; and it prolongs the life of the cutters between regrinds. All very laudable reasons to my mind! With only about $\frac{1}{32}$ inch of clean metal to be removed, run the lathe at the slowest open, i.e., non-backgear, speed and with the slide just nipped to make its movement slightly stiff, slowly feed the work past the cutter. If the lathe saddle can be locked to the bed, do so; we do not want any movement into or away from the cutter.

Reverse the soleplate to do the other side. Careful realignment of the work is necessary so that the two machined faces are as parallel to each other as possible and $\frac{9}{16}$ inch thick. The $\frac{3}{4}$ inch gap to take the bearings will be found to be cast very close to size, in my example less than 1/32 inch needed to be removed from each side using a square file. A four inch length of $\frac{3}{4}$ inch square bright mild steel bar was used as a gauge to ensure that the gaps were in line and to $\frac{7}{16}$ inch deep. Final fitting was done using the main bearing brasses.

Set the soleplate in position on the boxbed and spot through the four 17/64 inch holes. Open these up in the boxbed with a 13/64 inch drill and tap $\frac{1}{4}$ B.S.F. The tap must be started in the drilling machine to ensure that the studs will be truly perpendicular to the surface. Screw the four $\frac{1}{4}$ B.S.F. by $1\frac{1}{2}$ inch long studs into the boxbed, the end with the shorter thread going into the tapped holes, slide the soleplate over the studs and nut down.

MAIN BEARINGS—Item 5

These each comprise a lower and upper part and the first job is to face the mating surfaces. I did the lower brasses in the four-jaw chuck using a round nosed tool, the upper brasses were much closer to size and only required a few minutes' rubbing on a smooth flat file. The two lower brasses were then soldered together to make a cube of roughly one inch side and likewise the two upper brasses. The soldering was carried out using "Fryolux" Solder Paint. Each joint face was tinned, applying the paint with a small stick. While hot, the brasses were placed on top of each other, a final heat up with the bunsen type gas burner, some careful adjustments and the soldered blocks were left to cool down.

The block consisting of the two lower brasses was taken in hand first. It was machined in the four-jaw chuck to an overall size of 1 inch by 1 inch by $\frac{7}{8}$ inch, making sure that the joint came in the correct position. A $\frac{9}{16}$ inch end-mill in the self-centring chuck soon machined the channel, with the block held under the tool clamp, to the sizes shown on the drawing.

The gunmetal used for these castings machines beautifully at high speed but it is imperative that the cutting tools must have no rake angle so that there is no tendency for the cutter to snatch and pull into the work with disastrous results. This is particularly so with twist drills which already have a considerable rake angle due to the helix of the flutes.

The block comprising the two top brasses is held in the four-jaw chuck and the end faces turned to 1 inch diameter using a knife tool. Again see that this diameter is equally disposed about the joint line.

Unsolder the blocks and resweat them together in their correct pairs, carefully mark out the bore centres and mount in the four-jaw chuck with the centre mark running true.

The crankshaft mounted in the centred blocks for machining the crankpin. Note the work is wired to the catchplate driving pin.

Centre drill then progressively open out to $\frac{7}{16}$ inch diameter. Mark the bearing pairs so that they will henceforth be correctly mated and finally unsolder them and wipe off the excess solder while still molten. Mark out and drill the 13/64 inch holes in the top brasses and spot face.

Careful work is now called for with small files to fit

the bearing to the soleplate, finally spotting through and drilling 5/32 inch and tapping for the 2 BA studs.

For the final operation on the main-bearings, that of boring them out to size and truly in line, it was decided to use the method that was, no doubt, employed on the original engines, i.e. line-boring in situ, see Fig. 4. First a boring bar to mount between centres in the lathe was made from a 12 inch length of bright mild steel $\frac{3}{8}$ inch diameter bar. In its centre a $\frac{1}{8}$ inch hole was drilled through to take a cutter made from a broken centre drill, (we all make mistakes), clamped with a 5 BA grub screw. The boring was a slow process requiring care but worked beautifully, using a scrap piece of silver steel of correct diameter as a gauge. The important thing was never to alter the cross-slide setting when the boring bar was removed to check the bore diameter.

Now, I'll let you into a secret! Because my engine will have quite a bit of use, I made the crankshaft—and obviously the main bearings—to $\frac{9}{16}$ inch diameter. This gives some $12\frac{1}{2}\%$ greater bearing area which is always an advantage. If you decide to follow suit check your centering and setting out very carefully when it comes to making the crankshaft, which is our next job.

CRANKSHAFT—Item 1

Although often approached with trepidation, the machining of a crankshaft is not at all difficult as long as (a) it is not rushed; (b) all tools are really keen and well supported; (c) the various steps are methodically worked out and adhered to; (d) effective fixtures, etc., are prepared before they are needed, not as an afterthought.

Two accurate throwplates (Fig 5) are made first. The middle centre drilled position is so that we may turn the outside of the crankwebs. Now file flat the end faces of the crankshaft, carefully mark out, centre-punch and centre drill them. Next mount the crankshaft between centres and turn both ends to the required diameter. This is where the packing case wire comes in! Always wire the lathe carrier to the catchplate pin when turning between centres to keep everything secure. A side turning tool with a few degrees' top rake and a $\frac{1}{16}$ inch radius at the cutting tip will be ideal; include the web sides finishing the centre to $1\frac{5}{8}$ inch wide. Remove from the lathe and carefully file out the inner web surfaces leaving as little as possible to be removed by subsequent turning. Mount the throw plates on the crankshaft, not at the very ends, but as near to the centre of the shaft as the lathe catchplate and tailstock barrel will allow, (we want to keep the set-up as rigid as possible). With a parting type tool about $\frac{1}{8}$ inch wide with a rounded nose, and $1\frac{1}{4}$ inch overhang, (doesn't that unsupported overhang give one butterflies!), confidently start to machine, first the web sides then the crankpin. Run the lathe in backgear and do not exceed a 5 'thou' depth of cut. Give yourself plenty of time for this job and put a small piece of white card on the cross-slide so that you can see the tool approaching the webs at the end of each pass. With the crankpin to size, remount in the second pair of centre holes and turn the periphery of the crank webs. This is not according to how the drawing is detailed, but I prefer it this way!

The crankshaft may now be assembled in the main bearings and if our work has had the care which a set of Stuart Turner castings merits, we can spin the shaft and imagine it eventually revolving under the powerful thrust of steam. Another secret, I could not resist the temptation to see that large flywheel revolve so went ahead and machined and fitted it. In this form it illustrates page 5; I'll deal with its machining and fitting next.

FLYWHEEL—Item 21

There is a choice of flywheels for this engine. I presume that, like me, you have chosen the large spoked wheel, its proportions are so right for this particular style of steam engine, where speeds are rarely particularly high.

Machining will be carried out completely on the faceplate and we must prepare for this by making four packing and clamping plates. They are made from bright mild steel 3 inches long by 1 inch wide, two are $\frac{1}{2}$ inch and two are say $\frac{5}{16}$ inch thick. A clearance hole is drilled in the centre of each piece to take the faceplate bolts as required to suit the faceplate slots. It is also sensible at this point to make the keyway cutting tool (Fig. 6) which will be required later.

With the boss towards the headstock, clamp the flywheel to the faceplate. Use a scribing block or lathe tool to set the inner surface of the flywheel rim to run true (see the comment about setting up castings for machining in Part 1). When all is firm and true, put the lathe in its slowest backgear speed and face across the boss and the edge of the flywheel rim. If the lathe is not very rigid, face the boss, then centre drill and use the tailstock to give more support.

Reverse the flywheel on the faceplate, now use the thinner plates as packing. Again set the casting to run as truly as possible and make sure that the slide-rest will come sufficiently back to allow the periphery of the rim to be machined. You can, if you wish, omit the packing between the faceplate and the flywheel spokes thus the previously machined side of the flywheel now bears directly on the faceplate. If you do this be careful not to over-tighten the bolts and crack the flywheel spokes. If your faceplate runs perfectly true—and this is by no means certain—this method may obviate the need for a final skim with the flywheel mounted on a true running mandrel, e.g. crankshaft.

Turn the outside of the flywheel rim, the side, and the boss noting that this protrudes $\frac{1}{4}$ inch from the surface of the rim. Running the lathe at a higher speed, centre drill, open out and ream or bore the boss to give a very close, almost wringing, fit on the crankshaft.

Now we come to the fixing of the flywheel to the crankshaft. On the detail drawings a keyway is shown in the flywheel and a piece of steel is supplied for a flywheel key, no keyway is shown in the crankshaft. With this in mind and with reference to British Standard 46 "Gib-head Square Taper Keys and Keyways," it would seem that we have three possibilities open to us. First, we can make the key to full size dimensions, basing it on British Standard Keyway (BSK) $\frac{1}{8}$ GST (Gib-head Square Taper). This is shown at Fig. 7(a). One problem is that the material supplied is only $\frac{1}{4}$ inch wide while the key requires material $\frac{9}{32}$ inch wide. Secondly, we may presume the design of this engine to be about one-sixth full size. Hence the crankshaft would scale full-size at 3 inches diameter. For this BS 46 quotes a $\frac{3}{4}$ inch key and Fig. 7(b) shows this to a scale of one-sixth full size. In either case, if we are very fussy, particular depths of keyway in hub and shaft will be required as specified. This brings us to the third alternative—which I chose—that is to cut a keyway in the hub as specified ($\frac{1}{16}$ inch deep), ditto in the shaft, then file up as good looking a key as possible making the proportions look reasonably correct.

Cutting the keyways themselves is a slow but not a difficult operation. For the crankshaft a $\frac{1}{8}$ inch wide parting tool on its side, racked to-and-fro, with the crankshaft between centres is an easy way. Work towards the headstock and let each pass of the tool start slightly nearer the headstock. In this way the curved run-out effect will be obtained which is more correct than the rounded end that would be the result of using an end-milling cutter. Any slight steps in the keyway run-out may be removed by needle files. For the flywheel hub a short "boring" tool (Fig. 6) is required. Again take very fine cuts, only a 'thou' or two at a time, read the depth from the cross-slide index.

Now mount the flywheel on the crankshaft or on a mandrel and set the assembly between centres in the lathe. With a keen tool and slow speed, take the lightest skim off the periphery and sides of the flywheel rim. With a higher speed just touch the sharp machined edges with a dead smooth file to remove the sharp corners and the flywheel is complete and, most important, running true.

STANDARD—Item 6

With a small planing or milling machine this item becomes fairly easy and if you possess such equipment you will certainly not need me to tell you how to make use of it. But if only a small lathe and "manual shapers" i.e. files, are to be used, then careful thought must be given to evolving a methodical sequence of operations.

Careful checking of the casting showed that very little would need to be removed from the various surfaces to bring them to size, excluding of course the $\frac{5}{8}$ inch wide by $\frac{1}{8}$ inch deep channel for the crosshead. On a couple of occasions I have seen criticisms of Stuart Turner castings, in other publications, to the effect that they had insufficient machining allowance. This seems to me to show the writer's lack of ability to set up a casting for machining, or perhaps it is patience that is lacking. Castings which need large machining allowances before one finds decent metal are a criticism of the foundry and this is not the case

Standard clamped under the tool post for machining the crosshead channel.

with Stuart Turner. Now back to work!

I carefully filed and drawfiled flat and to size the vertical face of the casting which was checked with relation to the lower and upper horizontal bolting surfaces. When this vertical surface was satisfactory, the two afore-mentioned horizontal portions were machined using a flycutter (Fig. 8) held in the four-jaw chuck, with the Standard very carefully clamped to the lathe cross-slide (Fig. 9). Slow speed, a fine feed and I repeat, very careful aligning of the casting is all that is necessary. Do not rush, if you don't get it done today, it will be accomplished even more successfully tomorrow!

The two 13/64 inch and the two 17/64 inch holes at the top and bottom respectively were marked out and drilled. By the way, I felt that the bottom holes were too close to the centre web at $\frac{7}{8}$ inch centres, so increased this to $\frac{15}{16}$ inch. It allows greater clearance for the $\frac{1}{4}$ inch BSF nuts. Needless to say, the same modification is necessary on the Soleplate.

The final major operation on this component was to machine the $\frac{5}{8}$ inch by $\frac{1}{8}$ inch channel for the crosshead. To effect this, the Standard was clamped, with suitable packing, under the tool clamp. At first I was a bit doubtful, but with adjustable packing under the overhanging ends and the cross-slide nipped so that it would not snatch under the action of the cut, it all worked most successfully. As usual, I spent half-an-hour with a square file removing a fair amount of the metal from the channel, with the result that only three or four passes were necessary across the milling cutter to achieve the desired result. Although the cutter was a commercial slot drill, because of a slight lack of truth in the three-jaw chuck in which it was held, the channel was slightly over $\frac{5}{8}$ inch wide. This does not matter as the crosshead was made to fit.

GUIDE PLATES—Item 23

After a deal of awkward machining, a simple bit of benchwork will make a welcome change, and should be completed very quickly.

Saw the length of $\frac{1}{2}$ inch by $\frac{1}{16}$ inch bright mild steel

supplied to make these in half and drawfile one edge of each piece to get it nice and square. Mark out the hole positions, starting with the centre one. Centre punch and pilot drill the holes. Use the plates to position and spot through the holes on to the Standard, bearing in mind that the inside edges of the guide plates must be $\frac{1}{16}$ inch further in than the channel sides, thus leaving a $\frac{1}{2}$ inch wide space for the crosshead. Drill the holes in the Standard and tap 5 BA. Then open out the holes in the plates with a No. 30 drill.

Screw the plates to the Standard and file the outer edges down flush with the sides of the casting. Mark out and file the top and bottom of the Guide Plates to suit the Standard.

COLUMN—Item 11

This component, which years ago would have been a forging made by bumping up or heading the end of the bar, now appears to be made by friction welding. The initial operation is to face and drill the ends as all the turning will be done between centres in the lathe. With the work mounted in this way and with the lathe running at a fair speed, turn all over to about $\frac{1}{32}$ inch above finished sizes. Now set the tailstock over, (away from you if the Column foot is at the tailstock end, vice versa if at the headstock end). Check, by running the lathe tool up and down to see that a suitable taper will be achieved. Micrometer accuracy is not required here, but it should look about right. Now turn the taper, finishing with a very light cut at a fine feed.

If the tailstock of your lathe will not set over for taper turning, or you do not wish to alter it, turn this complete portion of the column to $\frac{7}{16}$ inch diameter. Then starting approximately 1 inch from the foot of the column, turn each one inch length to a little less in diameter as you proceed up the column. This will give you a series of steps which must be removed with file and emery cloth to get a nicely tapered column. The turning of the collar and $\frac{1}{4}$ inch diameter portion are simple, but of course the tailstock must be set back to turn parallel. I have a length of mild steel bar, centred at both ends, which I use for resetting and checking my lathes after taper turning.

The foot of the column has its edges turned to give them a nicely finished appearance and with the lathe driver wired to the catchplate pin it is well worth screwcutting the $\frac{1}{4}$ inch BSF thread at the top of the column to get it really square. This need not be completely cut but can be finished with a die. Trim to length and slightly round the end of the thread. Mark out, centre punch and drill the two 13/64 inch holes in the column foot, file to width and another component is finished.

BOTTOM CYLINDER COVER—Item 3

This item is not only a part of the engine structure, but is also the first of the cylinder components which we have to tackle. Start by giving it a clean, all over, with a file to remove parting line flashes or other excrescences. In particular, excess metal may be found joining the periphery of the flange to the boss which takes the column top. Removing this with a file will save considerably on machining.

Mount the casting by its rim in, preferably, the four-jaw chuck and set to run true with the lower surface of the casting outwards. Face and turn the chucking spigot so that it will afford a true location. If you feel that extra support is advisable, centre drill the spigot and use the tailstock. The seating to take the top of the Standard is now faced. This cannot be done with the lathe running,

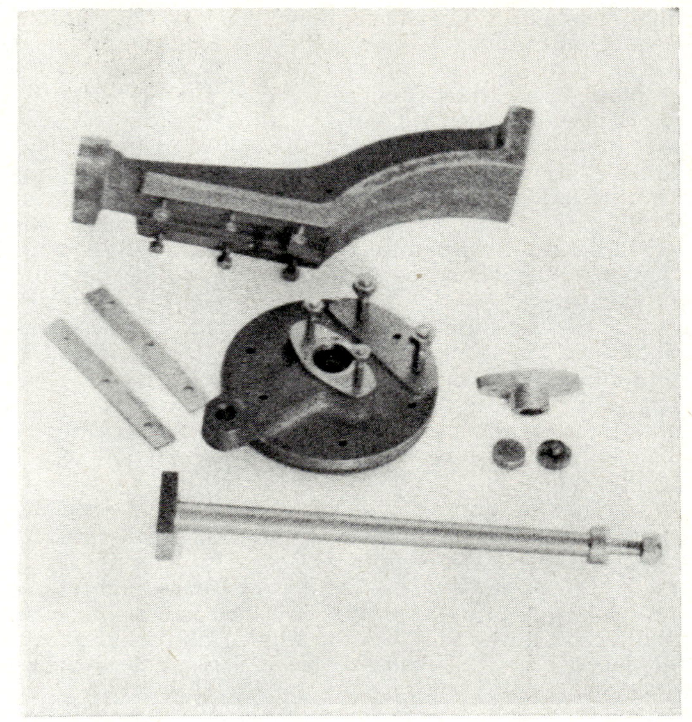

Standard, column and bottom cylinder cover. The filing buttons used are seen in front of the piston rod gland.

but by swinging the chuck back and forward through about half a revolution with your left hand while your right hand slowly feeds the cross-slide across. The seating should finish about $\frac{1}{16}$ inch higher than the unmachined surface of the casting.

Now reverse the casting in the chuck gripping by the turned spigot. This will only be about $\frac{3}{8}$ inch diameter so it is advisable to lightly skim the middle of the casting, centre drill and support with the tail-stock centre. Turn the outside of the cover to just under $3\frac{1}{16}$ inches diameter, to allow for a neat fit of the cylinder lagging which on measuring my sample appears to be of 26g. (0·0196 inch). During this operation finish the top side of the boss for the column.

The casting is now faced to the correct thickness and the $\frac{1}{32}$ inch high cylinder bore location is turned. If means of accurately measuring this are available this may be finished to 2 inches diameter, alternatively turn oversize, say $2\frac{1}{32}$ inches diameter or slightly more. The end of the cylinder may then be opened out to accurately locate on this spigot.

In practice, the ends of the cylinder bore and the covers often have this oversize location to allow for subsequent rebores. On engines which are to have much use this is a highly sensible idea but it is necessary to check on the availability of oversize rings and their size before such a course is worthwhile. Finally, with a sharp pointed vee tool just outline a $2\frac{7}{16}$ inches diameter circle on the cylinder cover flange. This will be of great help when marking out the six 13/64 inch holes for fixing the cover to the cylinder.

With the turning complete, centre drill the cylinder cover and open out progressively to 23/64 inch diameter. Work carefully, as this is getting close to the spigot diameter. Reverse again in the chuck, set the 23/64 inch hole to run true and bore out to $\frac{9}{16}$ inch diameter by $\frac{3}{8}$ inch deep. If a counterbore is available, or perhaps made, it will be useful in this operation. If not, accurate setting of the 23/64 inch hole and a small boring tool are the necessary requisites.

The last operation on this component is to mark out, centre punch and drill all the holes with the exception of the two 4 BA holes for the gland and the other two 4 BA holes for the valve rod guide (Item 29). Note that the $\frac{9}{32}$ inch hole for the column is at an angle of 8°. Be especially careful when the drill is breaking through on the underside that it does not snatch. A perfect example of the value of always drilling a pilot hole first and progressively working up to finished size.

We have just time for a little more, and a simple bit of work on the gunmetal Cylinder Gland will be a welcome change from steel and cast iron.

PISTON ROD GLAND—Item 14

You will notice from the detail drawings that there is a total of four 'elliptical' glands and flanges on this engine. They are all more or less of different sizes, but have in common the fact that in every case the ends are of $\frac{1}{4}$ inch radius with the centre coincident with the centre of the holding down studs or bolts. So that we get these glands looking really smart and to save considerable marking out, make a pair of filing buttons. These are of silver steel, $\frac{1}{2}$ inch diameter by $\frac{1}{8}$ inch thick with a pip $\frac{1}{8}$ inch diameter and $\frac{1}{16}$ inch long to locate in the holes. Harden them fully by quenching them in water from cherry red.

Mount the Gland casting in the four-jaw chuck and set the boss to run as truly as possible. Face the end of the boss and centre drill, steady with the tailstock centre and turn the boss to $\frac{9}{16}$ inch diameter, a close but easy fit in the bottom cylinder cover. Also face the flange. Now drill out the centre hole and ream to $\frac{11}{32}$ inch. Whether you machine the 120° chamfer or leave the gland end square is a matter of choice, personally I leave them square. Reverse the gland in the chuck and face to a thickness of $\frac{3}{16}$ inch.

Now mark out and centre punch the hole centres, drill $\frac{1}{8}$ inch and file each end to shape between the filing buttons. Blend the centre portion smoothly to the radiused ends and drawfile to a fine finish with a dead smooth file.

Slide the gland into the bottom cylinder cover and spot through into the cover, drill and tap 4 BA. Open out the holes in the gland with a No. 26 drill and assemble gland to cover using two 4 BA by $\frac{7}{8}$ inch long studs and nuts.

The various parts now to hand may be assembled. If you wish to be really fussy, make up stud boxes to drive the studs tightly home, or be like me and use two nuts locked together for this purpose. Don't use the nuts supplied as we want to keep them for the final assembly with their corners undamaged. The nut at the top of the column comes to rest on an inclined surface. Bad engineering practice and it looks paltry. Two choices are open to us, either file the boss so that the nut has a flat surface on which to rest, or make a taper washer—a device widely used on structural steelwork.

Being more of a steam enthusiast than a model engineer, I am always particularly interested to see small steam engines put to a real job of work rather than residing in a glass case. In this respect, I was pleased to read, in a steam journal, of a gentleman on the South Coast who for some years successfully powered a sizeable steam launch with a No. 1 engine. If readers know of other Stuart Turner engines working for their living, especially if the examples are out of the ordinary, I, and I am sure other members, would like to hear of them.

The great interest now developing in the so called 'soft' or 'low' impact technology characterised by Schumacher's 'Small is Beautiful' philosophy, has opened many peoples' eyes to the value of small steam power plants fuelled by virtually anything from highly expensive oil to the lowly country 'pancakes'! I must however, not allow my mind dwell on this, to me, fascinating subject, but rather help the beginner with his first set of castings to get into the game.

CONNECTING ROD—Item 12

I am not very sure whether I may have had the wrong casting in my particular set, but compared with the rest, there was quite a lot of metal to remove from this item. I suppose I should have contacted Stuart Turner's and checked this, but it was a wet Sunday, cosy in the workshop, the lathe humming merrily, and I just felt in the mood to make a good job of machining a con-rod. So if I spent more time than I need have machining a No. 1 rod from, perhaps, a No. 5A casting, then (a) its my own fault, and (b) I thoroughly enjoyed it!

The first operation was to file each end of the rod, then carefully mark out and centre drill. Using the centre drill holes as references, the excess metal was then marked out and sawn away. Do not use hacksaw blades with teeth which are too fine, it slows down the work. I use 18 teeth per inch blades for most work. One can really see the blade move through the metal. Particularly so when cutting out $\frac{1}{4}$ inch plate for boiler making.

The excess pieces of metal removed from the con-rod were not wasted. The rod is a malleable iron casting, and as my work is connected with materials, I decided to see what the microstructure looked like. Specimens for microphotography were prepared and, you can see what the material you are working looks like when magnified 100 times (see page 36).

The head of the rod was filed to about $\frac{1}{32}$ inch oversize all over and likewise, the foot. It was then mounted, between centres, in the lathe. Using a round nosed tool, the foot was machined to $1\frac{7}{16}$ inches diameter and the stem parallel and to $\frac{13}{32}$ inch diameter. The tool was replaced by a knife tool and the foot spigot turned to $\frac{5}{16}$ inch diameter, finishing the foot to $\frac{1}{4}$ inch thick.

Now the tailstock was set over for taper turning the stem, and this was carried out, again using the round nosed tool to get a smooth radius at each end.

The rod was removed from the lathe and the head marked out for the $\frac{9}{32}$ inch reamed hole and the $\frac{9}{16}$ inch wide gap, Fig. 10. The $\frac{9}{16}$ inch gap must be worked first. I did this by carefully and progressively drilling up to $\frac{1}{2}$ inch diameter. The con-rod was then clamped to the lathe faceplate, all alignments being as truly set as possible, and the $\frac{1}{2}$ inch hole bored out to $\frac{9}{16}$ inch diameter.

The accuracy and free running of the engine will largely depend on the alignments of the various parts of the connecting rod with both the crankshaft and the crosshead, so it is worth while taking some trouble with this part of the proceedings.

The remainder of the gap was sawn and filed away and the outer surfaces filed down to $\frac{7}{8}$ inch wide with the gap centrally disposed between them. Do not let this regular use of a file worry you. The iron prepared by Stuart Turners for their castings is so beautifully soft and even textured, that with an 8 inch medium and a 6 inch smooth file the removal of metal and the finishing of the surfaces becomes a real pleasure.

It is very important that the $\frac{9}{32}$ inch reamed hole must

be accurately parallel in both planes to the crankshaft. Do not rely on the drilling machine, unless you know it to be really accurate, for this operation. Rather set the rod up in the lathe, either in the four-jaw chuck, if big enough, or alternatively clamp it to the face-plate with judicious use of packing. I employed the latter technique, mounting the rod on a piece of 2 inch by $\frac{1}{4}$ inch bright mild steel in the manner shown in Fig. 11. This meant that the rod, accurately and firmly clamped to the sub-plate, could be easily adjusted on the faceplate to centre the hole position without upsetting the packing. The use of such 'sub-plate' techniques is invaluable in setting up components without convenient locating surfaces.

Centre drill the hole, drill and open out to 17/64-inch and ream, using a slow speed, to $\frac{9}{32}$ inch. In fact, if you wish, this is a situation where an increase in size, say to $\frac{5}{16}$ inch would not come amiss, especially if, like me, you can get hold of a $\frac{5}{16}$ inch reamer and not a $\frac{9}{32}$ inch one!

The rounding of the top end of the rod calls for a couple of $\frac{5}{8}$ inch diameter filing buttons. These are very useful for such jobs, and so easy to make that it is certainly not worth while attempting to do without them. A friend of mine built a 1 inch scale traction engine, with *all* gears hand filed, using filing jigs. Each was composed of a few teeth and radial strip for centre location. Not a job I would care to attempt.

The final job is to mark out and drill the two 13/64 inch holes for the big-end bolts in the foot of the rod. The drawing calls for them to be spot-faced. This is rather a difficult operation requiring an extra long or extension to the spot-facing cutter. I got over the problem by making the radius, shown as $\frac{1}{4}$ inch on the drawing, slightly less, so that the nuts would rest on a flat surface.

CONNECTING ROD BEARING—Item 15

You may have faced and soldered together the two gunmetal castings, which constitute this item, earlier on; when we sweated the main bearings together. If not, do it now. The mating surfaces need only light skimming in the lathe and the use of solder paint to sweat them together. Set them in the four-jaw chuck and face them to size, turning a $\frac{7}{8}$ inch diameter facing on each side. At one setting drill and ream, or bore, to a close running fit on the crankshaft, with as good a finish as possible. Throughout all this machining, it is important that the joint line is truly maintained on the centre-line. When the top surface is being machined, bore out the $\frac{5}{16}$ inch recess to accept the rod locating spigot.

With all the lathework complete, mark out the two bolt holes at 1 inch centres. Drill and open out to 13/64 inch diameter and spot face. On the underside of the bearing, very lightly centre with the smallest size centre drill.

Now assemble the bearing to the rod and bolt up firmly. Do not use the bolts obtained with the castings, but rather use a couple of odd bolts which do not matter if they perhaps get a nick from the turning tool during the next operation. Mount the connecting rod assembly in the lathe with the top of the rod held in the four-jaw chuck. Use two small packing pieces to bridge the gaps and do not be so fierce, when clamping the jaws, that the sides are squeezed in. Support the big-end on the tailstock centre and with light cuts machine the gunmetal bearing to match the connecting rod foot. This should be $1\frac{7}{16}$ inch diameter or thereabouts with the centre portion somewhat less, say $1\frac{3}{8}$ inches.

If you have, or can get hold of, a length of rod to go through the bearing bore accurately, and another length to go through the crosshead bolt hole, you can then sight

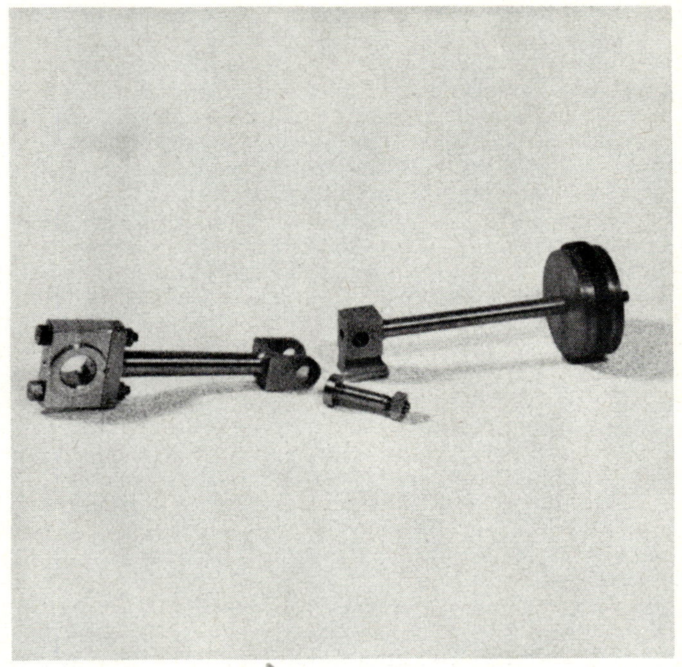

Connecting rod and piston components.

the parallelism of the two rods and thus assess the accuracy of your work. Even suitable size drills would do if nothing else is available.

Should any discrepancies exist, now is the time to rectify them by such means and wiles as may be necessary!

PISTON ROD AND CROSSHEAD—Item 7

Again, a malleable iron casting is supplied for this component. Clean over all with a file and trim to about $\frac{1}{8}$ inch over finished length. Accurately mark out the centres at each end and drill with a small centre drill. This will probably be most conveniently done in the drilling machine, in which case, use a trysquare to ensure that the rod is perpendicular when the drilling takes place.

Mount now, the rod between centres in the lathe, with the crosshead at the headstock end. You may find that there will be some difficulty in fitting a carrier to take a drive from the catchplate, so drill and tap a $\frac{5}{16}$ inch BSF hole in the catchplate sufficiently close to the centre-line to ensure that a driving pin, screwed into this hole, will make contact with the crosshead. Do not make this driving pin too long. As usual, wire the driving pin and driven portion together so that we have a firm contact.

With everything satisfactory, and a keen, round nosed tool, turn the piston rod down to $\frac{3}{8}$ inch diameter—not $\frac{11}{32}$ inch—but $\frac{3}{8}$ inch. Finish as accurately as possible to size. If you have a 0-1 inch micrometer, use it. If not, check your caliper over a $\frac{3}{8}$ inch drill or reamer, or piece of silver steel drill rod.

Swing the lathe tool over to an angle of about 30° to the centre line of the lathe and machine the upper face of the crosshead, checking it from the piston end of the rod. The $\frac{3}{4}$ inch radius will have a curve to it, but this can be levelled out by judicious use of a round file.

Now reverse the work between centres and face the end of the cross-head to finished length following the dimensions on the drawing.

To machine the rest of the crosshead, the lathe faceplate was set up and the piston rod clamped to it with suitable packing and clamps. I hope you will now see the reason for turning the rod to $\frac{3}{8}$ inch rather than $\frac{11}{32}$ inch diameter. It is simply, that the packing needed under the rod as work progresses, is of more standard sizes, and also, we can clamp against the rod without worrying about any damage to it.

Regarding clamps used on the lathe, I find that many of those sold as accessories are much too clumsy. I prefer them made from say, $\frac{3}{4}$ inch by $\frac{1}{4}$ inch bright mild steel bar and held with $\frac{5}{16}$ inch BSF bolts. A $\frac{1}{4}$ inch BSF hole at one end and a tinful of $\frac{1}{4}$ inch BSF round-head screws saves messing about with bits and pieces of packing.

The process of machining the crosshead, by a sort of simplified metal carving technique, is much too long-winded to detail step by step, but basically, it is as follows.

Clamp the rod to the faceplate with suitable packing. Face the crosshead portion until it is $\frac{1}{8}$ inch above the surface of the $\frac{3}{8}$ inch diameter rod. Using odd pieces of bright mild steel as gauges, between work and tool, will give you greater accuracy here, and more simply than using a rule. Then feed the tool in a further $\frac{1}{32}$ inch and machine the $\frac{9}{16}$ inch part, making sure to miss the $\frac{5}{8}$ inch wide foot. This is where the odd 'dummy run' is useful, just to make sure that the tool will not foul those parts which have to be left proud.

Now turn the rod over, and carry out a repeat operation on the other side, using this time a piece of $\frac{1}{8}$ inch packing under the rod and a piece of $\frac{1}{32}$ inch sheet under part of the crosshead. If the gap in your lathe will permit the piston rod to be swung about the crosshead bolt centre, mark out its position, set up on the face-plate, and drill and ream this $\frac{9}{32}$ inch hole, thus ensuring its true alignment with the rod. If the lathe will not permit the rod to be swung, still mount it on the faceplate but then transfer the faceplate to the drilling machine table and do the drilling and reaming operation there. Frequently turn the faceplate through 90° to cancel out any error. Actually, this is where a threaded end, or attachment, for the tailstock barrel can be jolly useful for operations of this type.

Try the crosshead in the Standard (Item 6) and see if it slides freely, but without sloppiness, in the channel. If any metal needs to be removed, take it equally from each side with a fine file. The top of the crosshead slider will need to be reduced to make it $\frac{1}{8}$ inch thick. This can be easily done with a 6 inch square file, until the crosshead slides smoothly, up and down, under the guide plates.

When you can stop admiring this silky fit, return the rod to the lathe, mounting it between centres, crosshead at headstock end. With a really sharp round nosed tool, turn down the rod to its finished diameter of $\frac{11}{32}$ inch. This has got to be right. Diameter, not only accurate, but also parallel throughout the whole length of $3\frac{1}{4}$ inches and as good a surface finish as possible. Nothing will cause gland packing to wear and leak steam faster than a rod with turning marks on it to cut and roughen the packing. So take the trouble to get it right.

With a knife tool, reduce the end to $\frac{1}{4}$ inch diameter and screwcut in the lathe to 26 threads per inch. The depth of thread is 0·025 inch and the tool angle, being Whitworth, is 55°. If you are using any other thread, as, for example, our American friends, then check accordingly. The thread can of course, be partly cut in the lathe, and finished using a die guided by the tailstock. However you do it, check frequently with the nut, as we want a firm fit between screw and nut. Any loosening of this particular nut, in service, could be mighty nasty! Finally, make sure that the centres still remain at each end of the rod, even if this means leaving the threaded end overlength

for the time being. We will require this facility if we wish to finish machine the piston between centres.

PISTON—Item 20

If you have means of accurately measuring in the 2 inch range, this item may be at least, partly machined now as we have just completed the piston rod. If not, then I suggest that you leave it until after the cylinder has been machined, so that it may be checked using the cylinder as a gauge.

Set it up in the lathe chuck with the chucking spigot outwards. Use the four-jaw so that you can get the casting to run as truly as possible. Face the piston, leaving only a little to be subsequently removed from the inner face. Turn the chucking spigot to clean up. Reverse and grip by the chucking piece in the three-jaw chuck. With a knife tool turn the recess for the piston ring. Check the $\frac{1}{8}$ inch length by using the piston ring supplied. Then turn the outer diameter to slightly over 2 inches diameter. Centre drill, drill 13/64 inch and tap $\frac{1}{4}$ BSF. As it is advantageous to keep any reciprocating part of an engine as light as possible, you may consider it worth while to recess out the inside of the piston as shown in Fig. 13, especially as the piston is made in two halves.

Repeat the above operations on the second part of the piston. No doubt you will, in fact, have worked them together. Fit them, together with the two piston rings to the piston rod and nut up tightly, checking that the rings are free to rotate with ease but not loosely. If satisfactory, remove the rings. I found they could be sprung off, but if you are in any doubt, dissemble the piston to remove them. Now reassemble and mount the piston rod, complete with piston between centre in the lathe. With a fine cut to ensure a good finish, turn the outside of the piston to finished size. You will read many suggestions as to what the clearance between piston and cylinder should be, with numerous erudite reasons given. Personally, I think that there are all sorts of variations to this theme, depending on size, materials, type of rings, running temperature and speed, whether saturated or superheated steam is used, lubrication and so on. My advice in the present case, is to aim for as close a sliding fit as possible, with the best surface finish. For this reason, I suggest that you may prefer to leave this particular operation until after the cylinder has been machined.

CROSSHEAD BOLT—Item 13

This particular piece goes under a number of names, but Crosshead Bolt it is to Stuarts, so Crosshead Bolt it will remain!

The material supplied is of $\frac{5}{8}$ inch diameter bright mild steel and of more than ample length, so it becomes a straight-forward bit of chuck work. A knife tool, so that the shoulders can be formed square, will do all the work. Aim for as good a finish as possible on the $\frac{9}{32}$ inch diameter (stone a slight radius at the tool point), and check it with the crosshead, to get a really close fit. The $\frac{1}{4}$ inch BSF thread should be done in the lathe, using a button die.

The final job is to fit the $\frac{1}{16}$ inch diameter snug key, whose purpose is to keep the bolt from turning as the nut is tightened or slackened and when working. Use a No. 53 drill to make the hole for a short piece of $\frac{1}{16}$ inch diameter silver steel. If your drill is rather short, drill the hole from the outside as I did. Tap the key through and file flush. Of course, before the key is driven in, use the hole in the bolt to spot through on the connecting rod. Drill the hole in the con-rod No. 52, which is clearance for $\frac{1}{16}$ inch.

Assembly of the items made so far will make things

Assembly of engine so far.

begin to look like a steam engine. When you assemble, do not tighten any bolt or nut more than finger tight, giving everything room to move. This is the purpose of the clearance holes at the points of location. Slowly turn the engine over, via the flywheel, checking the freedom of the crosshead and piston rod as they move up and down. Then slowly tighten the nuts, etc., starting with the connecting rod bearing bolts and working upwards. Include the studs between the bottom of the standard and the soleplate; and the top of the standard and bottom cylinder cover. The second anything nips, turn back one flat of the nut hexagon and make a written note to ease that particular point. Like me, I am sure you will find that, if you have worked to accurate rule measurement made frequent use of trysquare to ensure squareness, and tapped holes for studs in the drilling machine, you will be highly satisfied with your endeavours. There will, of course, be some closeness in the crank shaft and con-rod bearings, but this at this stage, is to be desired.

There is a tremor of excitement when one reaches the cylinder machining stage in any steam project. It is as though all the previous work has just been a prologue to this, the main single component. In actual fact, in terms of cost of casting and amount and accuracy of work entailed, the cylinder body probably fully deserves that extra care which we feel we should lavish upon it.

CYLINDER—Item 8

I never fail to experience a feeling of great satisfaction when I look at and handle a Stuart Turner cylinder casting. They are works of art. Beautifully proportioned, delicate cast-in steam and exhaust ways and metal which is a delight to work. For one to whom "the peace which passeth understanding" is found alone at bench and lathe, the prospect of settling down to a few hours of interesting work is delightful!

While you enjoy handling the No. 1 cylinder casting, notice that there are three possible surfaces which we might use as a datum for the machining of this component. These are the two cylinder ends and the port face. I chose the better of the cylinder ends and filed it flat and reasonably true, keeping it as square as possible with the cylinder axis. I had decided to perform all the cylinder machining operations on the faceplate of the small $3\frac{1}{2}$ inch lathe.

With suitable packing (actually, it only required odd slips of paper), the cylinder was clamped, by the prepared end, to the lathe faceplate. The casting was set to run as truly as possible by checking on the outside wall of the cylinder—not the flanges or the bore—but the outer wall, with due allowance for the port face.

With the lathe on the middle back gear speed, about one hundred revolutions per minute, the end of the cylinder was faced. The amount removed aimed at keeping the steam and exhaust ports equally positioned in the length of the cylinder block. After facing the end of the cylinder, set up a side turning tool and turn the cylinder flange to 3 inches diameter. This will include machining some off the end of the port face. A portion of the bosses for the drain cocks will be left unmachined and these are most easily done by hand filing. A few minutes work will bring the boss down to blend with the turned surface.

Reverse the cylinder casting on the faceplate and clamp it on some packing. Strips of $\frac{1}{8}$ inch by $\frac{1}{2}$ inch bright mild steel about 1 inch long will be admirable. They should be positioned under the clamps and serve to allow the boring tool to clear the cylinder bore without fouling the faceplate. The casting must be located so that the previously

Boring the cylinder

machined cylinder flange runs true. Now face this second flange but not down to length. It is better to finish finally to length after the cylinder has been bored. Turn the out-side of the flange to 3 inches diameter, reducing the end of the port face as before.

We are now come to the cylinder boring stage. The whole secret of success in this operation is to use as large a boring bar as the lathe toolpost and cylinder will accom-modate and mount it so that it has the minimum overhang. I have recently been machining some 3 inch bore cylinders and the boring bar which I was using in my 5 inch lathe was 1½ inches square with ½ inch diameter tool-bits. Round section tool-bits are better than square, because (a) the holder is easier to make and (b) the cutter may be rotated in its hole to set it at its best cutting position. Clamping should be with socket head grub screws.

Grinding and setting up the tool for boring is not always easy to the beginner, but a substantial boring bar will go a long way to ensuring success. Grind the tool bit to the shape shown, with about 5° top rake and just sufficient clearance to clear the cored cylinder bore. Use the slowest backgear speed and the finest feed. Do not be greedy with the initial cuts. The cored cylinder bore may not be concentric with the outside of the cylinder and only a portion of the surface may be being machined to start with.

To give some sort of example, you could easily find that perhaps half a dozen or more passes are required to remove the bulk of the metal, followed by three or four traverses with a re-sharpened tool to remove the last few thousandths of an inch and obtain the required degree of accuracy and finish. Let the final cut pass up and down the bore three or four times to remove any spring or deflection in the tool.

Aim for as good a finish as possible. Do not worry about whether one should leave toolmarks or should lap to finish—just aim for the best finish possible directly from the boring tool. In the final resharpening of the tool mentioned above, slightly radius the cutting edge with an oilstone slip or the oilstone on which you sharpen your woodworking chisels, etc.

Use a piston ring to gauge the accuracy of the bore. When it will slide in, using the piston to keep it square, and when you can just see a tiny slit of light in the ring gap when it is held up to the light, you are there! Theoretically, the gap should be calculated to suit the running temperature of the engine and measured with feeler gauges.

If you feel that perhaps the gap is a wee bit wide do not let it worry you. If the engine is to grace the mantlepiece it will not matter and if it is going to work, you won't even notice the extra piece of coal or cup of water that is used. If the gap resembles a barn door forget about using piston rings and use graphited packing instead, in which case aim for the very best surface finish possible in the bore, regardless of size, within reason, of course.

Before removing the cylinder from the lathe, face the flange down to the overall length of $2\frac{7}{8}$ inches, then recess the mouth of the bore to fit the spigot on the lower cylinder cover. Finally, file down the second drain cock boss to blend in with the cylinder flange.

Machining of the port face may be carried out by mounting the cylinder on the lathe cross-slide and using a flycutter mounted in the chuck or by setting up the cylinder on the faceplate and facing in the normal way. Whichever method is chosen first clean up the surface with a file and when machining use a very fine feed for the final pass to get a really good surface. I chose the latter method and made up a support to carry the cylinder for this operation from two short pieces of 2 inch angle. It will also be useful to support the cylinder when drilling and tapping the port face, especially as your drill vice probably does not open to take 3 inches.

The long sides of the port face are cleaned up by end

Facing cylinder portface

milling with the cylinder on the vertical slide or carefully packed up to the correct height on the cross-slide. Alternatively recourse may be made to small files and plenty of patience. I advise, and used the former, as the oval facing must be accurately flat to receive the exhaust flange.

If you intend fitting them, and it is advisable, mark-out, drill and tap the two holes for the drain cocks. Use the drilling machine for tapping to ensure squareness. With a small three-square (triangular) file, make a groove about $\frac{1}{16}$ inch wide and $\frac{1}{32}$ inch deep from the drain holes to the cylinder ends to permit water access to the holes when the piston reaches the end of its stroke. After all, water is incompressible, at least it is until it reaches a famous manufacturer of hydraulic equipment!

While the file is to hand, carefully bevel the outer edges of the cylinder ends of the steam passages, towards the end of the cylinder. Do it neatly and without scratching the cylinder walls. We are just making sure that the steam will reach the piston heads "without let or hindrance."

Still a little more judicious work with fine files. Carefully check the "as cast" steam and exhaust ports with the sizes as given on the drawing and trim them to their correct dimensions. Actually, it is not strictly necessary in the case of the exhaust port or the inner edges of the inlet ports, but it is worth doing and will take only a few minutes with some needle files.

The remaining work is the drilling and tapping of a multitude of 2 BA and 4 BA holes, but this must wait until the parts from which these holes are located have been made.

TOP COVER—Item 4

This item is a simple piece of lathework. Mount in three or four jaw chuck with the chucking spigot outward and set the whole component to run truly. Face and turn the spigot to as large a diameter as possible. Then face the top of the cover, removing approximately half of the machining allowance of metal. Reverse in the chuck, gripping the spigot in the centre of the inside jaws (see sketch). This means that the previously machined top surface of the cover is firmly supported against the pro-

Machining cylinder top cover

truding chuck jaws as the rim of the cover, etc., is machined. You will find that this gives more support and results in less chatter.

With this set-up, the rest of the top cover may be machined without difficulty. Obtain a close but not tight fit for the $\frac{1}{32}$ inch spigot in the top of the cylinder and with a small boring tool machine out the recess and clearance for the piston head nut. With a vee-tool strike a thin line on a $2\frac{7}{16}$ inches diameter pitch circle to help in marking out the holes for the holding down bolts. Remove from the lathe, saw off the chucking spigot, return to the lathe and with tool and file (if necessary) neatly finish the cylinder head dome.

Mark out the six holes using dividers set to $1\frac{7}{32}$ inches, centre punch, pilot drill and open out to 13/64 inch.

VALVE CHEST—Item 19

The valve chest is completely four-jaw chuck work and by the time you have finished it you will be expert at setting up work truly in the independent chuck. Take no notice of the remarks found in many books on lathework for amateurs about using chalk, you will find that setting with the lathe tool as reference, will give far greater accuracy and be much quicker. Do not use a hide mallet or any other percussive tool in an attempt to knock the job into a state of truth! When you have seen what such treatment can do to your expensive lathe chucks, you will be forever cured of inflicting such brutality. Likewise with vice handles. Never attempt to force them with lengths of pipe, etc. If they will not tighten securely, take them apart and clean them or make the necessary adjustments which will allow them to tighten properly. Bear in mind that the manufacturer of the equipment provided them with a handle which would supply all the torque necessary to fully tighten them.

Set up the valve chest in the four-jaw chuck with reasonable accuracy and face the top and bottom to the required thickness of $\frac{3}{4}$ inch. Remove metal in such a manner as to ensure that the two bosses finish centrally on their respective sides. Using the inside of the steam chest as a reference, mark out the outside of the rectangular frame. With the chest set reasonably true in the four-jaw

Facing boss of valve chest

chuck, face down to the witness line on the two plain sides. At intervals during the machining of this item, use a small try-square to check that the various surfaces are perpendicular with each other. Also check the overall length and width with that of the cylinder port face. Not that it matters very much, but we will be much prouder of our handiwork if the two line up precisely with each other.

Now mount the steamchest in the chuck with the previously machined end face pressed against the chuck body and set so that the valve rod gland boss runs truly. Check the squareness of all surfaces from the chuck face with a try-square. When true, face the gland boss to the correct length, then machine the end of the chest to its overall length of $2\frac{3}{8}$ inches making the major axis of the boss $1\frac{3}{8}$ inches wide. Centre drill the boss, drill through 11/64 inch and ream $\frac{3}{16}$ inch, then open out with a $\frac{3}{8}$ inch drill and bore to $\frac{7}{16}$ inch diameter by $\frac{3}{8}$ inch deep.

Re-set the casting in the chuck with this time, the steam inlet boss outwards. Again set to run truly. Face the boss until the casting is $2\frac{1}{8}$ inch wide, then reduce the side of the chest until it finishes 2 inches wide with the boss $1\frac{3}{8}$ inch diameter on its major axis.

The drawing gives $1\frac{9}{16}$ inches as the length of the flange boss, but my method of forming these flanges calls for the sizes stated above—take your choice. Centre drill and open out to $\frac{5}{16}$ inch diameter, this is where the steam will be fighting to get in.

We have not yet made the steamchest cover, the valve-rod gland, or the steam inlet flange, but when they are to hand they will be used to spot the various clearance and tapped holes in the steam chest.

The eight number 26 holes for the cover require little comment except to suggest that as they are quite long, occasionally rotate the steamchest while drilling, so that there is no tendency for the drill to run to one side and also, to allow for any lack of squareness between the drilling machine table and its spindle.

With the valve-rod gland in position and clamped, spot through the holes for the studs. Remove the gland, finish drill the holes No. 32 i.e. tapping size for 4 BA.

Put a filing button (with a suitable sized pip) in the number 32 hole, and using a small smooth file with a safe edge, carefully file the ends of the boss to $\frac{1}{4}$ inch radius. The remainder of the boss may then be blended in with the two ends making an otherwise awkward shape, neat and professional looking. The technique, as you will have guessed, is obviously the same with the steam inlet boss, using the gunmetal flange as a drill jig. I must admit cheating on the above, in order to use the existing filing buttons. I used a $\frac{1}{8}$ inch drill as the 4 BA tapping drill for the studs. The studs were then screwed in with just a touch of 'Loctite.'

VALVE CHEST COVER—Item 18

As we have the four-jaw chuck set up, it seems sensible to machine this item so that the valve chest is complete. Both surfaces, and all the edges can be done, but be very careful to take light cuts when doing the ends because of the excessive overhang. Likewise, be careful with the depth of cut so that you neither obliterate the "S" nor get the recess lop-sided. I do not wish to appear sacreligious, but you may fancy removing this letter and substituting your own, cut from sheet metal and rivetted or sweated in place—useful having a name like Smith!

After the edges of the cover have been reduced to match the valve chest, lightly drawfile them to remove the turning marks, unless of course, you intend to paint them.

Mark out, centre punch, pilot drill and open out to No. 26, the eight stud holes, then use it to jig drill valve chest and cylinder port face as previously described.

If you intend to fit a displacement lubricator, catalogue page 34, item No. 185/1, now is the time to drill and tap a $\frac{1}{4}$ inch by 32 tpi hole in the centre of the side of the valve chest opposite the steam inlet. This will save having to dismantle the engine later on. It is well worth fitting this lubricator, not only for the obvious improvement which results in the running of the engine, but also because the existence of an oil film in the cylinder will protect it from rust—always a problem in an engine which is only run occasionally.

VALVE ROD GLAND—Item 17

Still using the four-jaw chuck, mount the gunmetal gland casting with the boss outwards, and set to run as truly as possible. Face and turn the boss to $\frac{7}{16}$ inch diameter, a close sliding fit in the valve chest. Face the flange and reduce the length of the boss to $\frac{5}{16}$ inch.

Centre drill the boss, drill 11/64 inch and ream $\frac{3}{16}$ inch. Then bore the mouth to an included angle of 120° or leave square, as desired.

We have had a long session of four-jaw chuck work and more to come, but for the present, change to the self-centring chuck and mount the gland, gripping by the boss. Face the flange to $\frac{3}{16}$ inch thick.

Remove from the chuck, and mark out the two stud holes at $\frac{7}{8}$ inch centres. Drill to $\frac{1}{8}$ inch and again, with our, oh! so useful, $\frac{1}{2}$ inch filing buttons, radius the flange ends. As usual, blend in the remainder of the flange to form a neat shape.

Fit the gland to the steam chest, and check their alignment by sliding the piece of $\frac{3}{16}$ inch diameter stainless steel through both gland and valve chest. The fit should be smooth with no hint of sticking. If not up to standard, try the gland the other way round, i.e. rotated through 180°. If still tight, choose the best position and pass the $\frac{3}{16}$ inch reamer through.

So that we may complete the drilling and tapping of the remaining holes on the cylinder and valve chest, we

Centre-drilling for valve-rod gland

might with advantage, continue by dealing with the machining of the steam inlet and exhaust flanges.

EXHAUST FLANGE—Item 26

Clean the gunmetal casting with a file and set it in the four-jaw chuck with the boss facing outwards. Very careful setting will be required because the outside of the boss has to finish $\frac{5}{8}$ inch diameter and as cast, it is almost to this size. Have sufficient of the flange proud of the chuck jaws so that it may be faced at the same setting.

After turning, centre drill, open out with a 29/64 inch drill and tap $\frac{1}{2}$ inch by 26 tpi. This is the British Standard Brass Thread as used for brass tube, etc. It is of Whitworth form and is in fact, the same as $\frac{1}{4}$ inch BSF. I find it invaluable for steam work, particularly boiler fittings, and use it in from $\frac{1}{4}$ inch to one inch sizes. One small advantage, which accrues from the fact that it uses the same thread form and size throughout, is that the tapping size may always be easily arrived at. It is simply 3/64 inch less than the thread diameter. Also when making up boiler fittings, the lathe can be kept set up for cutting 26 tpi regardless of work diameter.

Reverse the flange mounting by the boss in the three-jaw chuck and face the flange to $\frac{1}{8}$ inch thick. Mark out and drill the bolt holes $\frac{1}{8}$ inch at 1 inch centres. then with the filing buttons radius the ends and blend in the rest of the flange to match. Finally, open out the bolt holes with a No. 26 drill and use the flange to spot the 4 BA tapped holes on the cylinder exhaust outlet.

INLET FLANGE—Item 26

There is no need to set out in detail, the machining of this component, as it is a repeat performance of the previous item. In fact, if the two are worked together, it will save a lot of chopping and changing of lathe chucks, tools, etc. The only differences are, to drill the boss 21/64 inch and tap $\frac{3}{8}$ inch BSB (brass), and to drill the bolt holes at $\frac{7}{8}$ inch centres. Again, use this flange to spot the holes in the inlet boss cast on the side of the valve chest. With the filing buttons and a smooth flat file with a safe edge, the inlet boss may be shaped to match with its flange—of course, if you are feeling lazy, you can skip such 'gilding the lily' tactics!

VALVE ROD HEAD—Item 2

It is surprising to find such a large piece of mild steel provided from which to make this small item. In fact, I checked against the schedule of parts because I felt sure that there had been an error, only to find that the size supplied was in accordance with that listed. The $\frac{3}{8}$ inch dimension is in order, as this is the correct width, while $\frac{3}{4}$ inch, although over-long provides plenty for gripping in the lathe chuck. But to reduce $\frac{5}{8}$ inch thickness down to the required $\frac{9}{32}$ inch, seems to be work for the sake of it! However, the day was cold, and a session with hacksaw and file soon had the blood flowing—not literally! The steel was reduced to $\frac{5}{16}$ inch thick and a $\frac{1}{8}$ inch cross hole drilled at $\frac{3}{8}$ inch from one end.

The job was then mounted accurately in the four-jaw chuck and the end faced down so that it was $\frac{5}{16}$ inch from the centre of the previously drilled hole. The end was centre drilled, drilled and tapped 2 BA. Take care as the drill breaks through into the $\frac{1}{8}$ inch hole. Finish the bar end by radiusing with the edge of the lathe tool. Then remove from the chuck and with a $\frac{3}{8}$ inch diameter filing button (what would we do without them!), neatly radius the opposite end. Open out the $\frac{1}{8}$ inch hole to 9/64 inch, and reduce to $\frac{9}{32}$ inch thick by drawfiling on a smooth flat file. This latter action, may also be used to correct any lack of concentricity in the position of the tapped hole.

Finally, a comment on the 2 BA tapped hole. Before you perform this operation, carefully read and give thought to the remarks about threading the next component, which is the . . .

VALVE ROD—Item 27

Little work is actually needed on this $\frac{3}{16}$ inch diameter stainless steel rod apart from facing in the three-jaw chuck to $3\frac{3}{8}$ inches long and threading both ends. However, it is the apparently simple items which may cause the greatest trouble, and this could be one of them!

The reasons are two-fold. First, the material. Stainless steel can be awkward to work requiring keen tools, and how often do we regrind our dies! Secondly, a 2 BA thread (0·1850 inch outside diameter) is called for on a $\frac{3}{16}$ inch (0·1875 inch) diameter rod.

The best way to make a first-class job of this important piece, would be to set up truly in the four-jaw, or make a little $\frac{3}{16}$ inch bore split bush for the self-centring chuck, and mount the rod in that. Then screw-cut the threads, finishing if desired, with a die. The snag here, of course, is that a 2 BA thread has 31·4 tpi, not much help!

If you have $\frac{3}{16}$ inch fine thread, say 32 or 40 tpi, taps and dies, use them, screwcutting first. This was my procedure. If you must use 2 BA, take a couple of 'thou' off the ends of the rod with a smooth file in the lathe and then thread, using a tailstock die-holder.

All this may seem a lot of fuss, but an inaccurate or drunken thread on this component, will mean that we can never adjust the slide valve without it sticking and perhaps even lifting from the port face.

With both these items made, grip the valve rod head truly in the four-jaw chuck and the valve rod in the tailstock drill chuck. In this way we can screw them together with maximum accuracy of alignment.

VALVE ROD GUIDE—Item 29

This neat little gunmetal casting, calls for some crafty setting up in the four-jaw chuck. I find the $4\frac{1}{2}$ inch diameter

First operation on valve-rod guide

size light pattern just right for this sort of lathe "sculpturing."

After cleaning with small files, set up in the lathe with the flat flange outwards. Machine flat and to about $\frac{1}{8}$ inch thick. Always be very concerned with such shapes

Drilling and reaming valve-rod guide

to the casting of 1 inch, then centre drill, drill 11/64 inch and ream $\frac{3}{16}$ inch. Remove from the lathe and finish the top surface of the boss with a file.

Mark out the positions of the two fixing holes and drill them $\frac{1}{8}$ inch, then—guess what? Yes, that's right, use the $\frac{1}{2}$ inch filing buttons to neatly shape the flange. Finally, open out the $\frac{1}{8}$ inch holes with a No. 26 drill.

SLIDE VALVE—Item 16

The initial checking of the dimensions of this component revealed that fettling had been carried out over-enthusiastically with the result that the rear sliding face of the valve would finish at only $\frac{3}{16}$ inch wide instead of the required $\frac{1}{4}$ inch. In this respect, notice how important it is that the valve cavity be maintained in the centre of the valve face. However, with things as they were, there was no way I could maintain the necessary dimensional accuracy. Without any doubt Stuart Turner would have exchanged the casting, but if one is on a desert island and the last pigeon of the day has gone, one does not feel disposed to waste valuable machining time! So a small piece of $\frac{1}{4}$ inch by $\frac{1}{8}$ inch section brass, $1\frac{1}{8}$ inches long was wired to the cleaned up rear end of the valve, heated with the propane Sievert brazing torch and touched with silver solder. No more than ten minutes work and the casting had been saved and more important, I was able to get on with the job.

Using the four-jaw chuck, machine the face of the valve removing the minimum amount of metal so that the cavity is maintained at maxium depth. Also, machine the sides of the valve to $1\frac{1}{8}$ inches square, keeping the cavity central. In my particular casting, the cavity was very accurate in terms of length ($\frac{5}{8}$ inch), but slightly narrow on the $\frac{3}{4}$ inch width.

and constantly check the relationship of the surface being machined, with the remainder of the casting.

Rest the flange against the chuck face boss outwards and running true. Face the boss to give an overall length

The upper surfaces of the valve were then cleaned with a file and the position of the $\frac{7}{32}$ inch hole marked out. The casting was set up in the drill vice and this hole drilled. First, a small centre drill was used, so that the hole was started in the right position. Then the hole was opened up progressively to finished size. In addition to the centre drill, three sizes of drill were used to ensure that there would be little tendency for the hole to run out.

The $\frac{5}{32}$ inch wide slot, to take the valve nut, may be cut without difficulty, by clamping the valve under the lathe tool-post and traversing it past a $\frac{5}{32}$ inch slot drill—or drill ground flat—held in the lathe chuck. Instead, I measured the thickness of a 6 inch smooth, safe edge, flat file and found that, if used on edge, it would produce a slot pretty near to $\frac{5}{32}$ inch wide. Thus, I completed the slot in quicker time than it would have taken me to set up the job in the lathe.

However, I did eventually set up for end-milling in the lathe, because I decided to mill out the cavity to as deep as possible and to the full $\frac{3}{4}$ inch wide. Easy access of exhaust steam to the exhaust port, and so to atmosphere, results in a more free running engine. So, after carefully measuring the depth of the nut slot, etc., I milled out the cavity to a full $\frac{1}{8}$ inch deep. It may make little difference, but heck, I know it is done right!

A light rub of the valve face on some fine emery cloth on the surface plate and the slide valve—not the easiest shape to cope with—was finished.

VALVE NUT—Item 10

This tiny, but very important, link in the chain is simplicity itself to produce. There are, however, certain details, which, like most of the small bits and pieces in engineering assemblies, have got to be right.

Cylinder assembly

In the case of this nut, its thickness, nominally $\frac{5}{32}$ inch, must fit the slot in the valve easily but with no slop, a test of good fitting. Also, the 2 BA tapped hole must be square with the surface of the nut. If you have used some alternative thread for the valve rod (item 27) then obviously it must be used on the nut.

ECCENTRIC SHEAVE—Item 28

We must make certain that the slide valve, over which we have taken so much care, is opened and shut at the correct times and in sequence with the position of the piston. It is the duty of this particular item to see that this is so.

Clean the chucking spigot with a file and mount in the three-jaw chuck. Lightly face the boss and centre with the smallest size of centre drill. Support with the tailstock. With the lathe in back-gear, turn the sheave to $1\frac{3}{8}$ inches diameter. Remove from the chuck and mark out the bore centre to give an offset of $\frac{3}{16}$ inch. Now mount in the four-jaw chuck with this bore centre running true and centre drill, drill and bore or ream to form a close, push fit, on the crankshaft. Then turn the boss to $\frac{7}{8}$ inch diameter and reduce to $\frac{1}{4}$ inch long. Skip the remainder of these instructions until the strap is made.

After sawing off the chucking spigot, reverse the casting in the four-jaw chuck, gripping by the boss and set the sheave, previously turned to $1\frac{3}{8}$ inches diameter, to run true. Now finish turn the sheave to the sizes given. Aim at the best possible finish, so that a sweet running eccentric sheave and strap will result. Face the sheave to $\frac{5}{16}$ inch thick.

Mark out, drill and tap the 4 BA grub screw hole in the boss. Now, I may be blind, but I just could not find such a grubscrew in the multitude of setscrews, studs, nuts, etc., supplied. Even a study of the list of parts failed to show any reference to such an item—strange!

ECCENTRIC STRAP—Item 22

If, way back at the very beginning, you followed my advice, and sweated together, with soft solder, all those gunmetal parts that required to be machined as a pair, this item will have been waiting patiently for some notice to be taken of it.

Now set it up in the three-jaw chuck, gripping by the inside jaws in the bore. This by the way, will be a good test of the effectiveness of your soldering.

Skim the outer face, removing only sufficient to clean up overall. Change round the facing tool, so that it will machine the periphery of the casting and, with light cuts, turn the top edge, i.e. where the eccentric rod fits. Again, only remove sufficient metal to clean up this surface. My particular example finished at $1\frac{1}{4}$ inches radius.

Remove the casting from the lathe, and carefully clean up with files all the remaining unmachined surfaces. The shoulders, on which the heads and nuts of the 4BA clamping bolts will seat, should be dealt with by being reduced to $\frac{13}{16}$ inch wide. Now mark out, using the machined face for reference, the No. 26 holes and carefully drill them. Make absolutely sure that the holes go through square. If you are at all doubtful of the accuracy of the drill vice, clamp the casting to a small angle plate. Whichever method you employ, keep turning the strap round, as drilling proceeds, to try and cancel out any error which may exist.

While at the drilling machine, the surfaces can be spot faced with a $\frac{5}{16}$ inch diameter cutter.

Now mount the casting in the four jaw chuck with the unmachined side outermost. Set so that the outer surfaces, which will be left unmachined, run reasonably true. Face the side of the casting and finish, nominally, to $\frac{5}{16}$ inch thick. I say 'nominally' because, for appearance sake, we want the bolt holes, eventually, to lie in the middle of the thickness of the finished strap. Obviously, if necessary the eccentric sheave must be adjusted to match.

With a keen boring tool in the toolpost, bore out to

$1\frac{1}{4}$ inches diameter. Achieve a really fine finish. We do not want any excessive wear between the eccentric strap and the sheave.

In truth, how often are the eccentric straps of small steam engines adjusted for wear? If they are, by the usual method of filing some off the bolting faces, we get a bore which is no longer round and fits where it touches! Lately, on some engines, I have fitted non-adjustable straps. The sheaves were cast-iron, the straps from bright mild steel and a really close fit. I reckon I'll be kicking up daisies before they need adjusting, with the amount of work they get. On the other hand, if they really are going to be run day after day, fit ball races between sheave and strap, as on the 1899 steam car engine in my workshop. Of course, I'm obviously not referring here to the scale model type of engine, where the method used should conform to the original.

With a bit in the boring tool holder, which matches the rib to be machined on the eccentric sheave, bore out the recess in the strap. The strap may now be clamped to the vertical slide, or on a piece of angle, and bolted to a cross-slide, and set to the correct height to end-mill the $\frac{3}{8}$ inch wide recess for the lower end of the eccentric rod. Hold the cutter in the three-jaw chuck, and run the lathe at top speed. Take a light cut and read the end of the cut from the cross-slide index. If there is any tendency to chatter, nip up the slide-rest gib strips so that they are just slightly stiff. Don't overdo this, or you will not be able to feel the tool cutting.

Remove from the lathe, drill the 9/64 inch hole and after stamping some identification, so that reassembly will always be done correctly, unsweat the two parts of the strap. Clean off the excess solder, and with a fine file, deburr the two parts.

Return now to the eccentric sheave instructions and complete the work described in the last two paragraphs of that section, using the strap as a gauge.

Drill the oil hole in the top strap, as shown, and assemble strap to sheave with some thin machine oil and gently run in by hand, lightly nipping the bolts as you do so.

ECCENTRIC ROD—Item 24

With this malleable iron casting, we come to the final constructional item in the building of the engine. Although it may not be immediately obvious to a beginner, the greatest difficulty in this type of item is to ensure that the various holes and radii are concentric with each other. Accuracy of centre distances are, in general, not too difficult to achieve, but with holes which are not central the whole appearance is spoilt. The solitary rod in this engine will let you in gently to the time when you decide to fit the full Stephenson reversing gear, as, no doubt, you will.

Start by facing the back of the rod flat and true. This may be done by end milling, with the rod held in the vertical slide, or facing, with the rod in the four-jaw chuck—a rather less secure method of holding. Alternatively, like me, you may simply drawfile the back, using a smooth, flat file. Lightly clean up the edges, then file the top surface finishing at $\frac{5}{32}$ inch thick. But for the moment, leave the head at the full thickness.

Mark out and drill at the 4 BA hole locations. Make up the necessary filing buttons—you will eventually build up a most useful collection of these—and radius the ends, $\frac{3}{16}$ inch and $\frac{5}{32}$ inch respectively. Now mark out and finish file the edges, to blend neatly with the radiused ends.

Carefully mark out for the $\frac{9}{32}$ inch wide slot, using the back face of the rod as a datum. Drill progressively up to $\frac{9}{32}$ inch diameter. Then, in the lathe, mill out the slot with a

$\frac{1}{4}$ inch end mill and finish to width with a small smooth file. With the slot complete, file down the head, ensuring that the slot is centrally located. Then tap the 4 BA holes —I suggest in the drilling machine—and open out one side to 9/64 inch diameter.

Fit the rod to the eccentric strap, easing as necessary, but maintaining a reasonably tight fit, and screw together. If any tendency to slackness exists here, sweat with soft solder, or you may prefer to be more modern and apply a touch of 'Loctite.'

FINAL ASSEMBLY AND PAINTING

I am sure that every one of you building this engine will have a different technique of going about this last exciting bit of the work. In fact, you would disappoint me if I thought you had stuck religiously to each instruction given so far. I hope that you will have tried out your own ideas and methods, using these instructions as a guide and an indication of how, at least, one individual did the job.

If you have assembled the engine, and checked its smooth working, as construction proceeded, there should be no problems in achieving final accurate assembly. I dismantled my engine completely and laid all the parts out on sheets of paper on a table. Each surface, which it was intended to paint, was given a coat of undercoating, using an artist's type brush, about $\frac{1}{4}$ inch wide. The choice of undercoating is, of course, dictated by the choice of final colour. It was suggested to me that all the unmachined cast sufaces be filled, so that a mirror-like finish would be obtained, but it struck me that we were dealing with an example of steam engineering, not a heap of sheet steel from an automobile showroom. So it was a case of two undercoats and two top-coats picked out in red with polished bronze and steel and a brave show it makes. The paint

Temporary assembly. Stuart No. 1 Engine stands on bed of lathe on which it was made

used was Humbrol, dark green. I believe I am right in saying that the colour originally used by Stuart Turners on their products is described as 'bronze green.'

The fitting of the lagging was done before the cylinder was painted. The blue planished steel was first bent round a piece of 3 inches diameter bright steel bar. Be careful not to scratch the blued surface. In the future when you polish the engine, simply wipe over the surface of the lagging with a rag just moistened with a very clear and thin machine oil, don't rub so hard that you eventually rub through the oxide film on the surface of the lagging.

After bending the lagging sheet, it was sprung round the cylinder and carefully marked out for trimming to fit. Do not be tempted to cut to size with a pair of tinsnips. They are sure to deform the cut edges, and the evidence will always be visible. Either cut some distance away from the line, and finish with a dead smooth file, or use a miniature hacksaw. In either case, clamp the lagging between a couple of narrow strips of thin plywood with the line level with their edges. Drill and tap the necessary holes in lagging and cylinder for the 5 BA screws to hold the lagging. Fit the lagging and mark out the positions of the drain holes. It is advisable to clamp the sheet between two strips of mild steel when drilling these holes which will be $\frac{1}{4}$ inch diameter. This will ensure a clean edge.

The cylinder surface was undercoated and, when dry, built up with some asbestos sheet which had been just moistened with water to make it mouldable. The asbestos, at $\frac{1}{4}$ inch thick, is nowhere near thick enough to make any difference, as far as reducing any heat loss is concerned, but it provides a backing for the lagging sheet. I imagine that a substance like 'Polyfilla' could be used in the same way, although personally I have not yet tried it. In my case, the asbestos was moulded round the cylinder and left to set after which the lagging was permanently fixed.

The joint faces at each end of the cylinder, between port face and steam chest, and under the steamchest cover, each require an oiled paper gasket. This requires more work, but is much neater and cleaner than using a compound such as 'Osotite,' which tends to squeeze out and spread everywhere. Use stout brown paper and mark out the shape accurately especially the size and position of the holes. Cut out really carefully and then soak for ten minutes in a clean, thin, machine oil. Allow to drain before fitting so that you don't get oil dribbling over all the painted surfaces as you tighten up the nuts.

Use fine, graphited yarn packing in the piston rod and valve rod glands, but do not tighten down the gland nut more than finger tight. These glands must not cause any stiffness in the free sliding of the rods, or they will mop up a great deal of power. Under steam, they can be nipped up just sufficient to halt any leakage, but if frequent adjustment becomes necessary, carefully inspect the rod, you will almost certainly find that there is a rough patch that is abrading the packing. There is only one cure for this, that is to replace the rod and be more careful in handling it next time!

With the setting of the valve, which is the next procedure, we are getting very near the stage when we will be able to try the engine under steam for the first time.

To set the valve, remove the steam chest cover, and, with the eccentric locked to the crankshaft in any position, turn over the engine via the flywheel. Do not be too vicious, as the valve may be so badly out that it may clout the top or bottom end of the steam chest. Adjust the valve by taking out the bolt linking valve rod to eccentric rod and rotating the valve rod right or left hand as necessary, until the valve is opening the steam ports equally at both ends.

Now loosen the grubscrew locking the eccentric sheave to the crankshaft, and rotate the flywheel in the direction of travel, until the piston is at top dead centre. As you are probably facing the steamchest, the direction of rotation will be anti-clockwise. Now rotate the eccentric sheave in the same direction, i.e. anti-clockwise, until the valve reaches the top of its upward travel. Now, put on your spectacles, and as you slowly continue to turn the eccentric sheave, watch the valve start to travel downwards. As soon as the finest possible line of the steam-port shows stop, and lock the eccentric sheave grubscrew. Sit back and let the tension subside!

Next, continue to turn the engine over, in an anti-clockwise direction, using the flywheel. When the piston reaches the bottom of its stroke, the valve should have reached the bottom of its travel, be now rising, and have reached the point at which the fine line of the lower steam port is just showing.

The amount of port visible at the top and bottom dead centres mentioned above is called the 'lead' of the valve, and a little thought will show you that what is actually happening is that the valve is opening shortly before the piston reaches the end of its stroke, i.e. top or bottom dead centre. If this lead is not equal at both ends of the stroke, the valve can be adjusted slightly, up or down, to put matters right, by rotating the valve rod. That is, by screwing it up or down through the valve nut. But do realise that the 2 BA thread, between the rod and valve nut, has a pitch of 0·81 mm (nearly 0·032 inches), thus the minimum adjustment possible (half a turn of the valve rod) will move the valve up or down some 0·016 inches, i.e. 1/64 inch. So don't be too worried if you do not get it exactly right.

If you are miles out, the only thing to do is dismantle the valve and eccentric and carefully check your workmanship and the port sizes. I do earnestly advise that, as long as the components are to the sizes given on the drawings, and you have a reasonable valve setting, that you leave matters alone for the moment. The steam engine may be a simple, straightforward, converter of energy, but there are many ways by which its efficiency can be improved, or, for that matter, reduced, by fiddling about with valve and port sizes and eccentric throws. This engine can give you a life-time of interest as your knowledge of steam power grows and you try out various valve settings and modifications.

As you have drilled and tapped for the fitting of drain cocks, these can now be screwed in with, if necessary, a tiny smear of jointing compound. The section arrangement drawing No. 70079, shows them to be straight nose cocks (catalogue No. 196/2). However, I would suggest, especially in the interests of safety, that you do not fit this type but, instead, use union cocks (catalogue No. 200/2) and pipe the outlets to some safe place below the baseboard or into a tank. Steam, even at quite low pressures, is highly dangerous stuff, if treated without respect; and it is much too easy, with the original type, to receive a blast of steam or boiling water directed at you. I have personal experience of this! In addition, youngsters have an irresistible habit of twiddling knobs, and the damage would be done before you could stop them.

With the engine coupled to an air, or, much to be preferred, steam supply, and all bearings liberally oiled, you may sit back and revel in the sight and sound of the creation of your skill and patience running. But treat it with great respect until you get to know it, it has enormous power. As you open up or close down the supply, so the engine will speed up, or drop to a tick over no faster than your own heartbeat.

For me, this particular work has reached its end. For you, it has reached its beginning, because a vast field of interest in steam power now lies before you.

The finished engine was run in on the lathe; this method is preferable to using compressed air if the parts are rather stiff, besides which, you have got everything under much better control. When all is working smoothly and easily, the final job, painting, can be tackled. You will then have the pride and pleasure of seeing the fruits of your labour in all its glory; don't keep it under a glass case—make it work for its living!

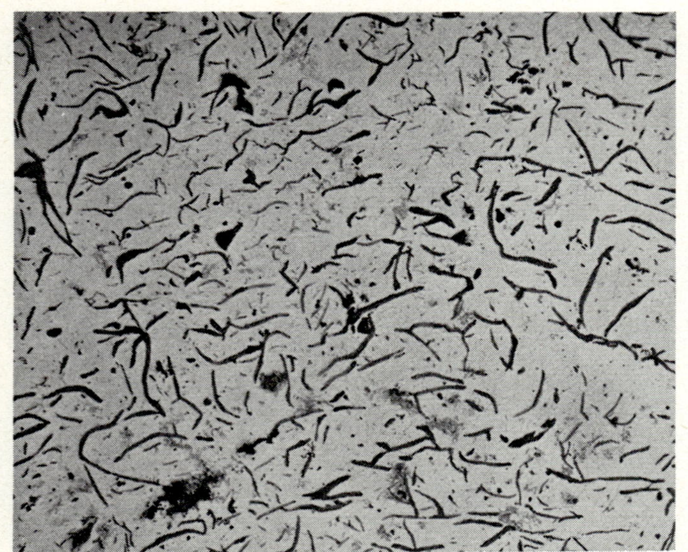 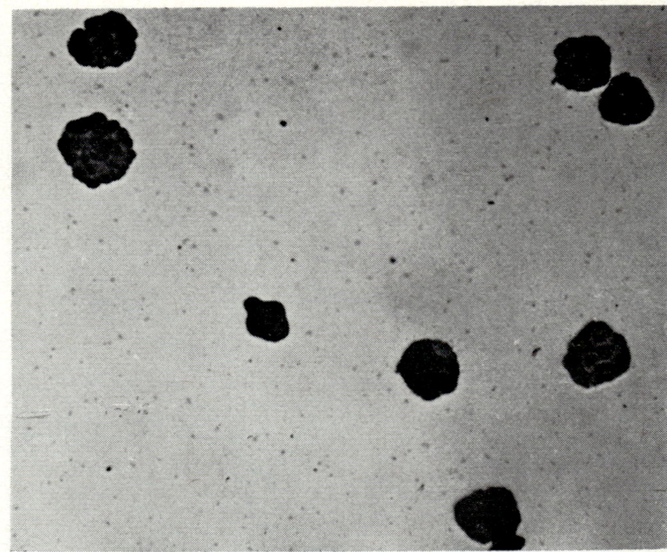

MICROSTRUCTURE OF CAST-IRON

One is not involved in the machining of metals for long before it becomes very apparent that cast-iron is rather brittle. You will have confirmation of this if you are unfortunate enough to drop an iron casting on a concrete floor or grip it badly in the vice!

The reason for this brittleness is seen in A which is of a plain, good quality cast-iron. The black streaks are flakes of soft graphite, the space occupied by the flakes each being, in effect, a minute crack in the structure of the iron. Any sudden blow on the casting may start a crack running through the iron from flake to flake resulting finally in breakage.

How is it therefore, that we find such items as the crankshaft, connecting rod and piston rod which need to have some degree of ductility, being supplied in the form of iron castings (malleable iron)? The second micrograph (B), which is of the specimen cut from the con-rod, shows how the brittleness has been considerably reduced and ductility increased, by causing the graphite to gather together into balls, containing a minimum of sharp corners from which cracks may start. This is one of a number of methods by which iron castings may be made more ductile.

The micrographs were taken with a hand-held camera, through the eyepiece of a microscope at a $\times 100$ magnification.

BUILDING THE No. 1. ENGINE.   Fig 1

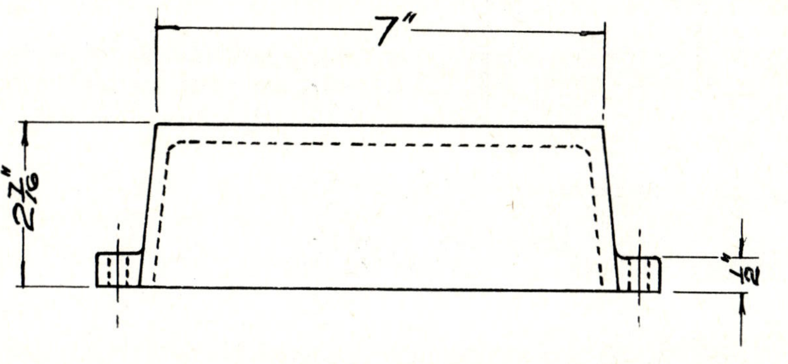

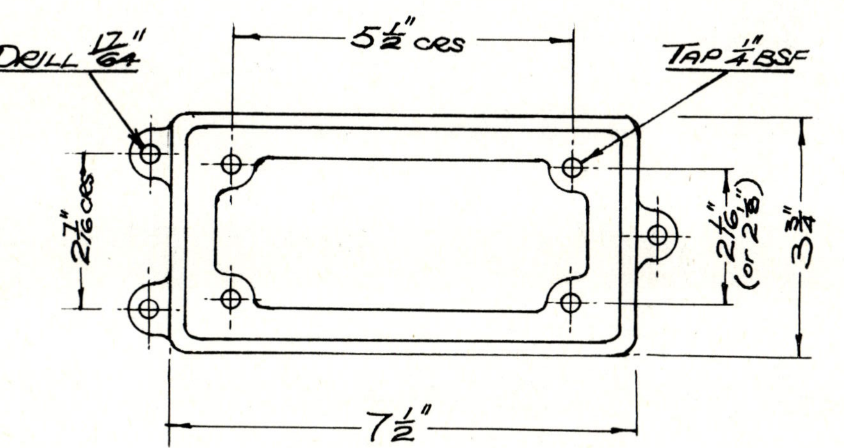

Check these dimensions with individual casting.

STUART TURNER No 1 ENGINE

BOX BED PART No. X2.

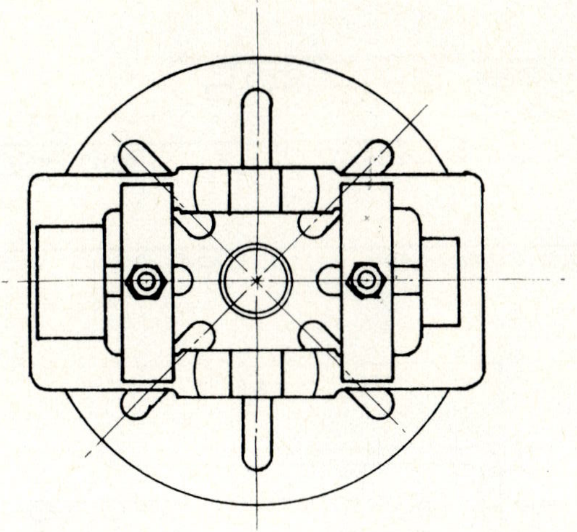

FIG 2. Soleplate clamped to faceplate.

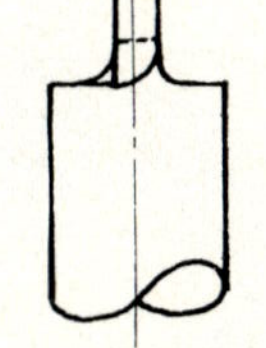

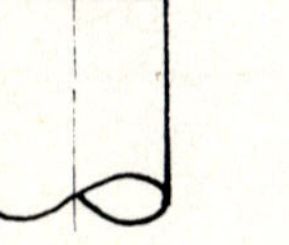

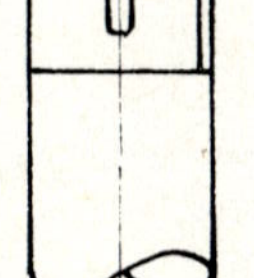

FIG 3. Milling Cutters.

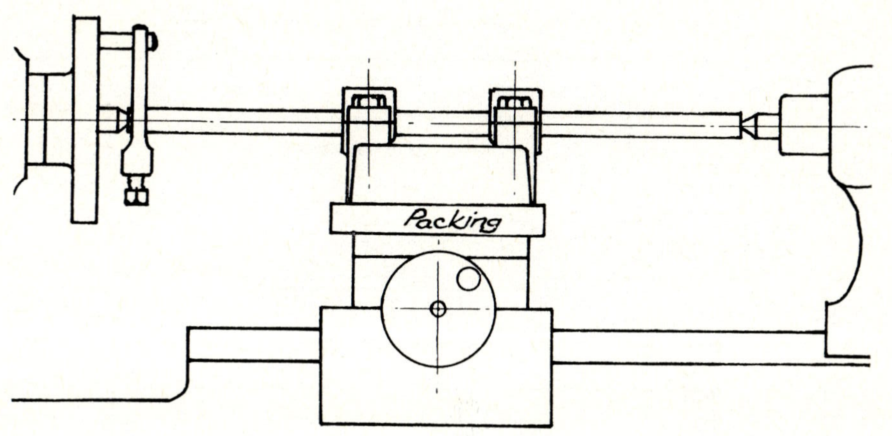

Fig 4. Line-boring of main-bearings.

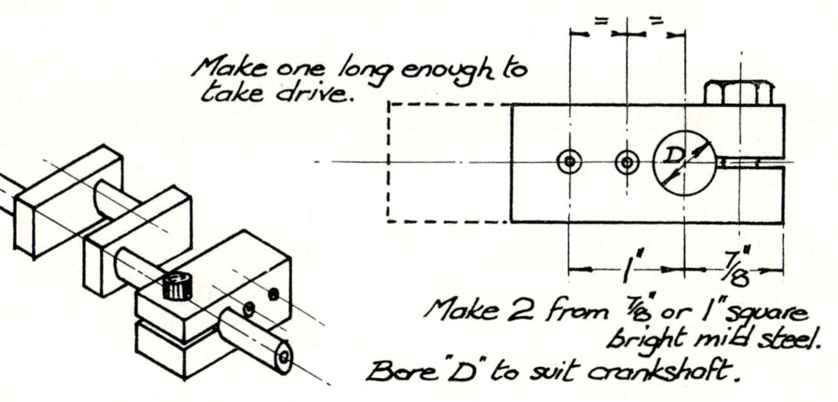

Fig 5. Throwplates for Crankshaft.

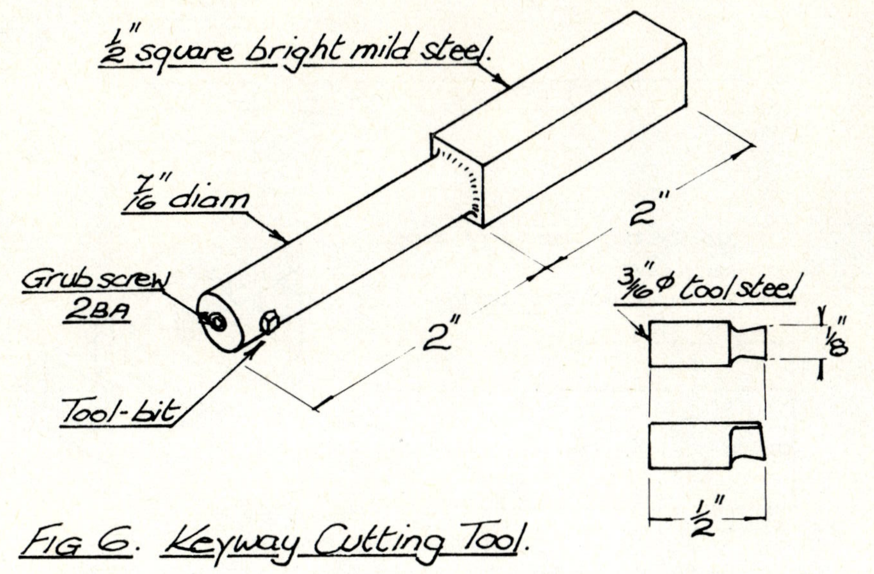

Fig 6. Keyway Cutting Tool.

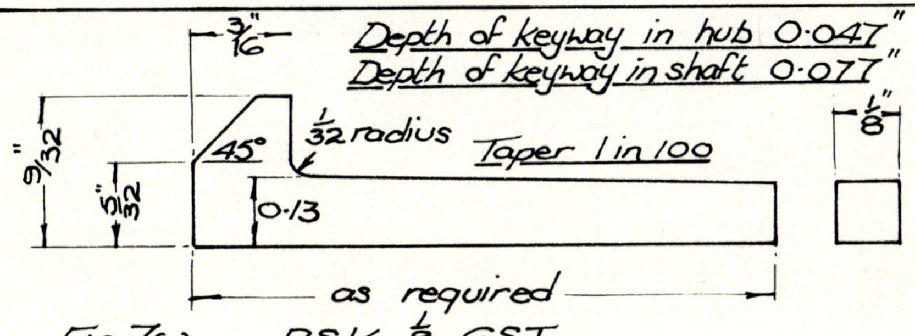

Fig 7(a)   BSK 1/8 GST.

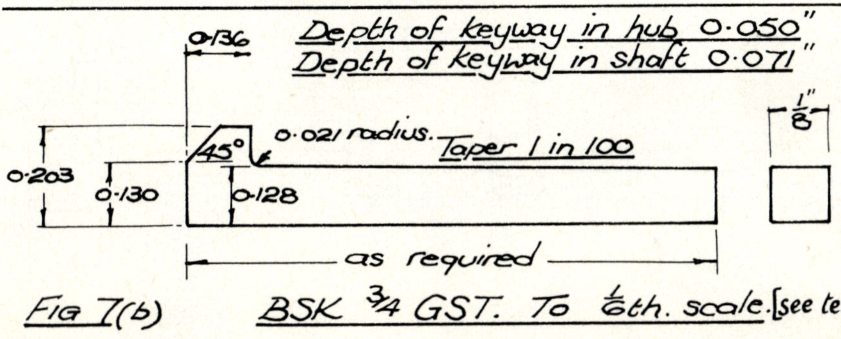

Fig 7(b)   BSK 3/4 GST. To 1/6th. scale. [see text]

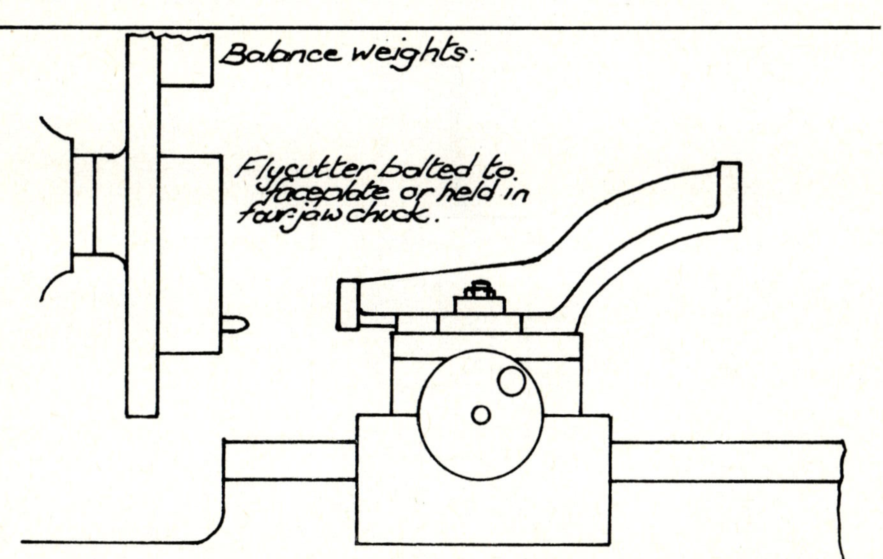

FIG 8. Useful form of flycutter for face-milling in a small lathe.

FIG 9. Machining Top & Bottom of Standard.

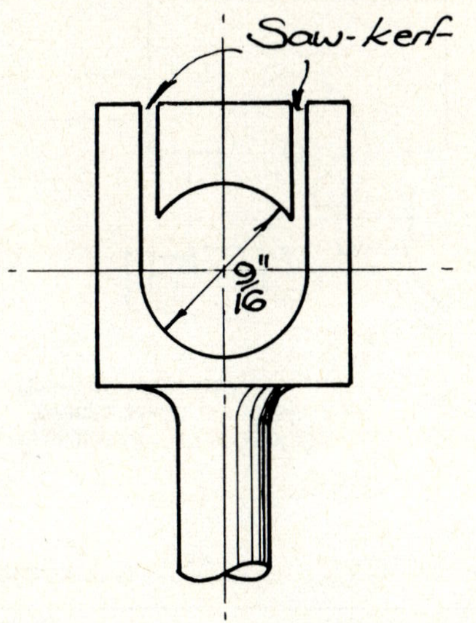

Fig. 10. Sawing out top of Connecting Rod.

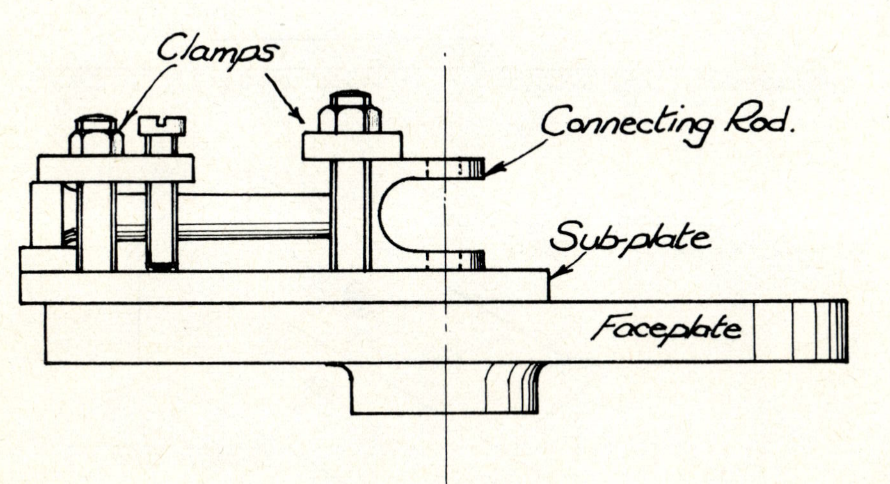

Fig. 11. Con-rod clamped to faceplate for boring.

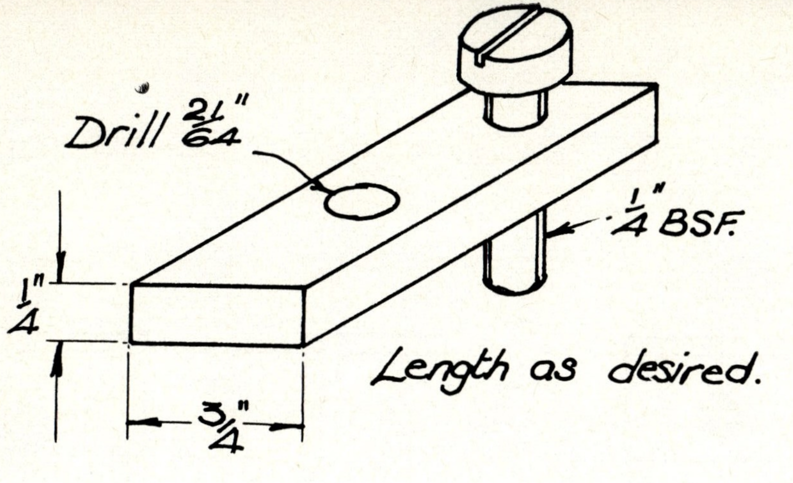

Fig. 12. Clamps.

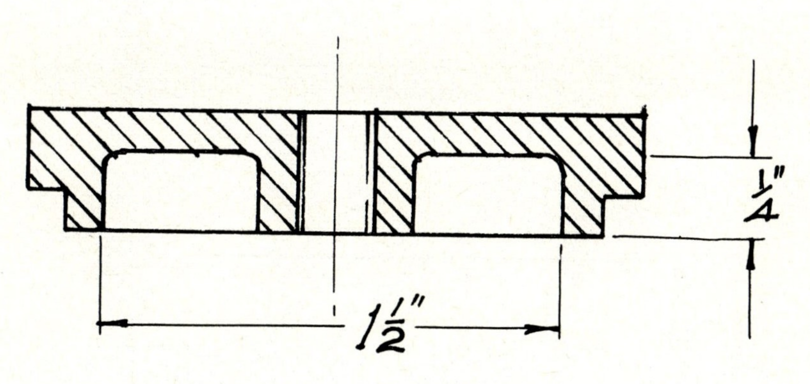

Fig. 13. Hollowed Piston.

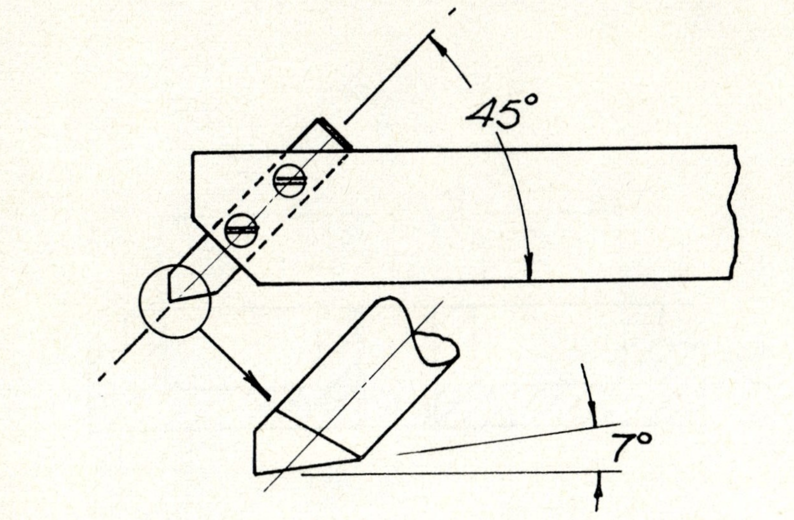

(14) Boring Bar and Cutter.

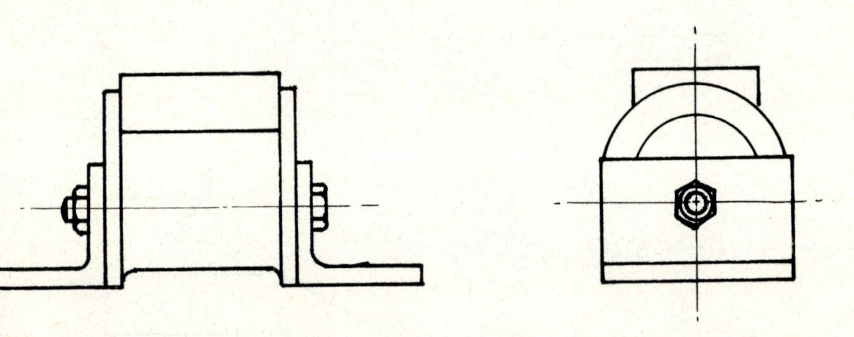

(15.) Brackets to support Cylinder.

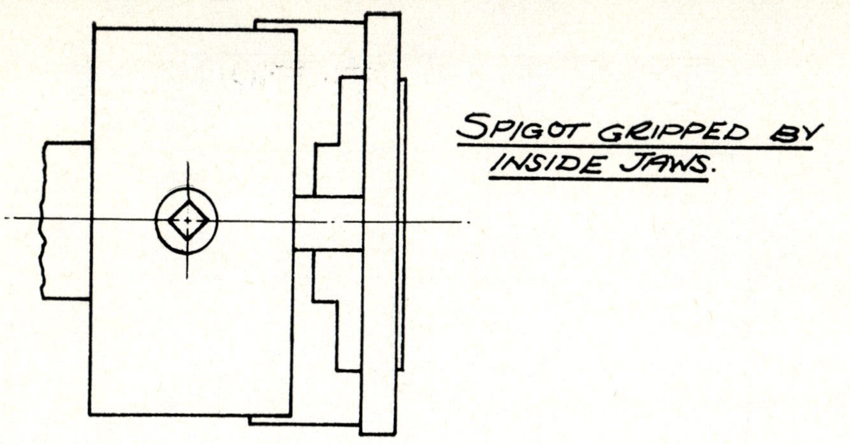

(16) Holding top cover in S.C. Chuck.

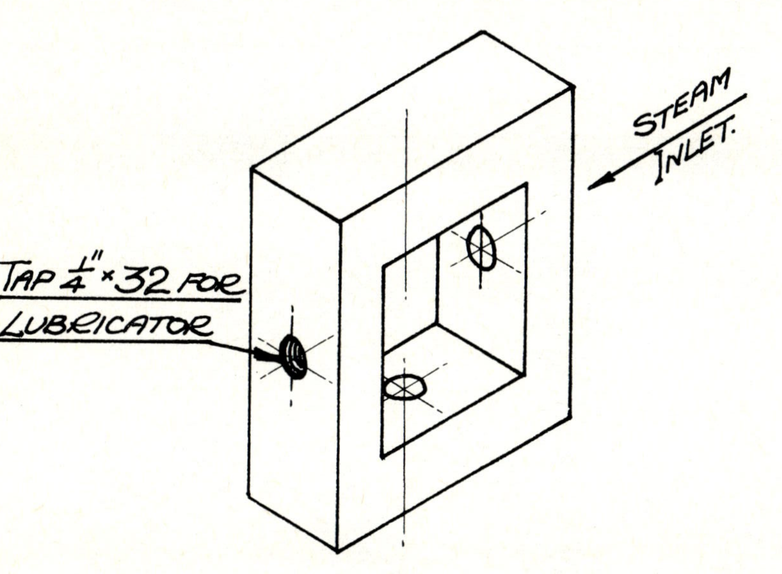

(17) Position of Displacement Lubricator.

45

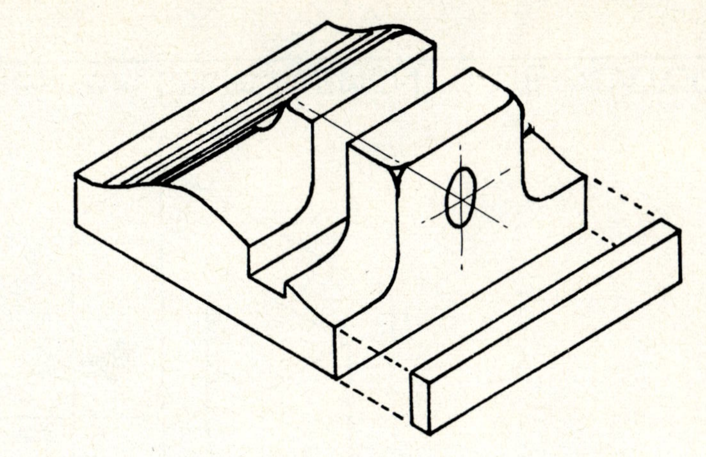

⑱ Addition to slide valve

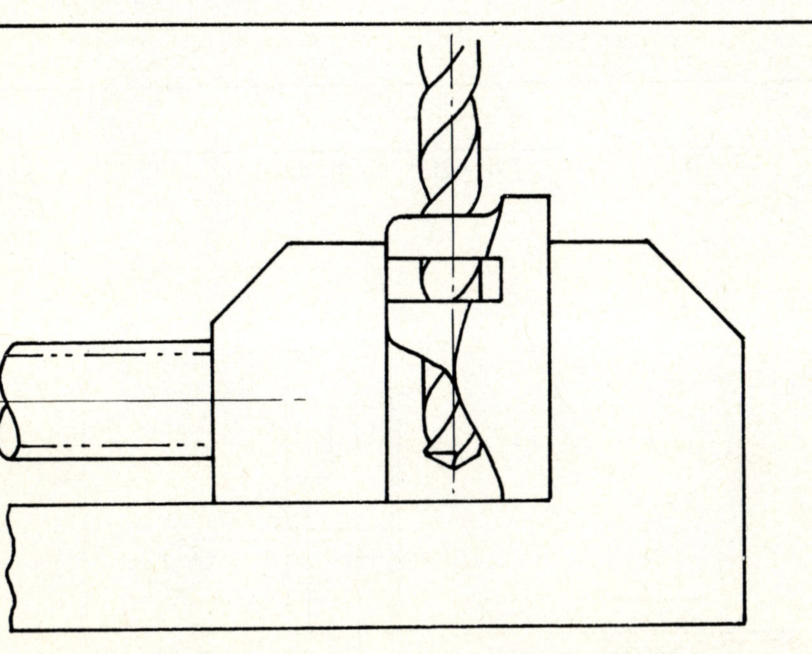

⑲ Drilling Slide Valve.

A LITTLE THOUGHT WILL SHOW THAT WITH THE ENGINE ROTATING IN THE DIRECTION SHOWN, THE FORCES IN THE PISTON ROD AND CONNECTING ROD, WILL RESULT IN A THRUST HAVING TO BE WITHSTOOD BY THE CROSSHEAD GUIDE. THE OPPOSITE DIRECTION OF ROTATION WOULD RESULT IN THE CROSSHEAD ATTEMPTING TO LIFT AWAY FROM THE SURFACE.

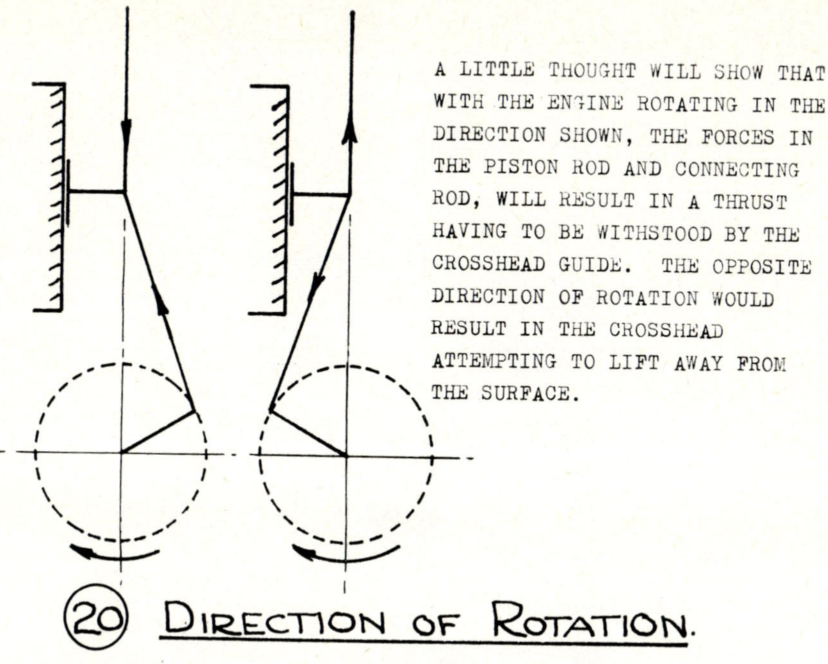

(20) DIRECTION OF ROTATION.

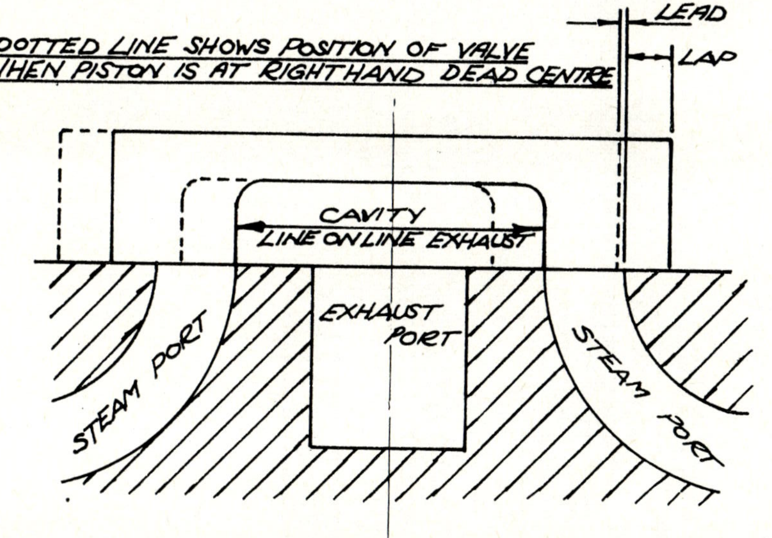

(21) DETAILS OF PORTS AND VALVE

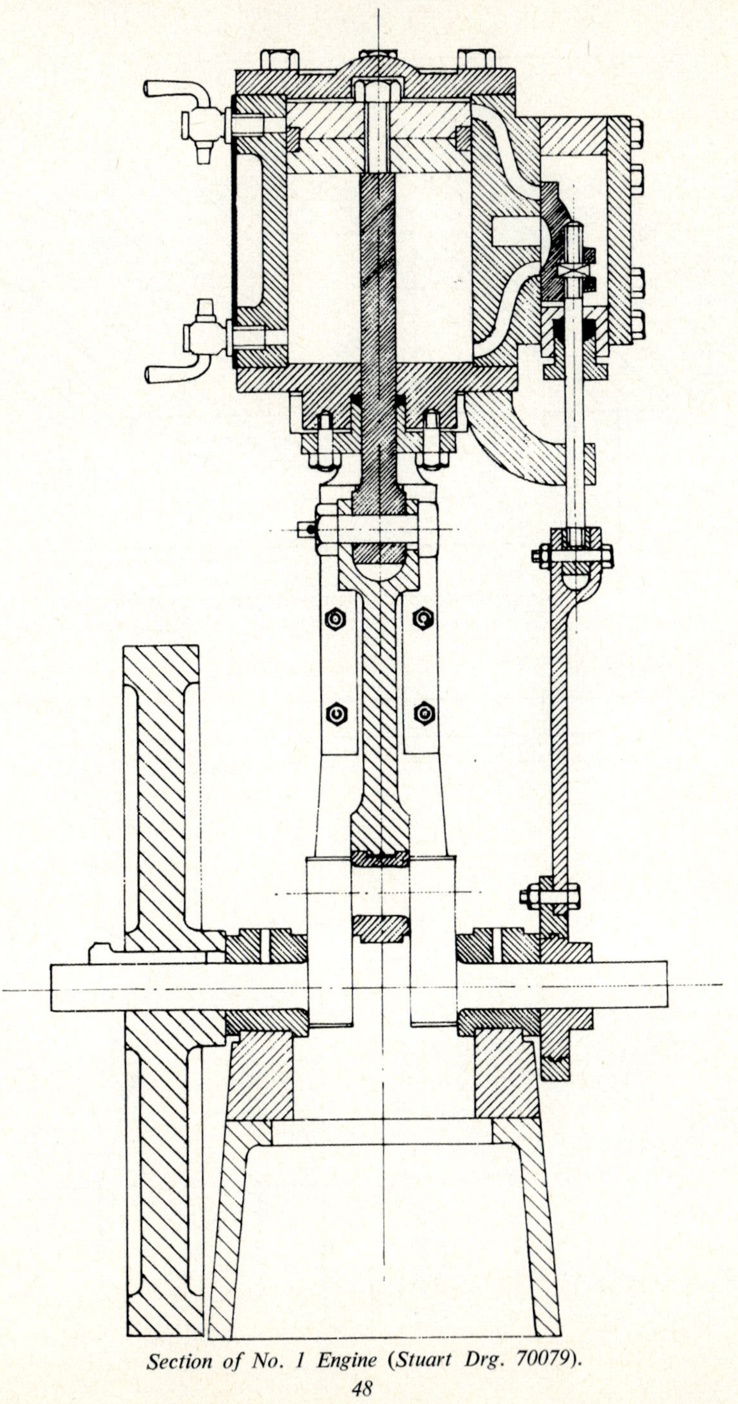

Section of No. 1 Engine (Stuart Drg. 70079).

48